Andrés Rey

Application of Crowdsourcing to Generate Datasets

Andrés Rey

Application of Crowdsourcing to Generate Datasets

of Public Procurement Processes

ScienciaScripts

Imprint
Any brand names and product names mentioned in this book are subject to trademark, brand or patent protection and are trademarks or registered trademarks of their respective holders. The use of brand names, product names, common names, trade names, product descriptions etc. even without a particular marking in this work is in no way to be construed to mean that such names may be regarded as unrestricted in respect of trademark and brand protection legislation and could thus be used by anyone.

Cover image: www.ingimage.com

This book is a translation from the original published under ISBN 978-613-9-07351-1.

Publisher:
Sciencia Scripts
is a trademark of
Dodo Books Indian Ocean Ltd. and OmniScriptum S.R.L publishing group

120 High Road, East Finchley, London, N2 9ED, United Kingdom
Str. Armeneasca 28/1, office 1, Chisinau MD-2012, Republic of Moldova, Europe
Managing Directors: Ieva Konstantinova, Victoria Ursu
info@omniscriptum.com

Printed at: see last page
ISBN: 978-620-8-38508-8

Contents

CHAPTER 1 2
CHAPTER 2 3
CHAPTER 3 5
CHAPTER 4 7
CHAPTER 5 8
CHAPTER 6 20
CHAPTER 7 21
CHAPTER 8 37
CHAPTER 9 54
CHAPTER 10 67
CONCLUSIONS 78
RECOMMENDATIONS AND FUTURE WORK 79
BIBLIOGRAPHY 80

CHAPTER 1

1. INTRDUCTION

The World Bank defines corruption as the exploitation of public power for private gain, highlighting its negative impact on the economic and social development of countries. The phenomenon is intertwined with a range of social, economic, legal, cultural and psychological problems, making it a complex challenge to address. Despite its complexity, transparency is seen as fundamental to combating it. In Colombia and other countries, e-government and free access to public information laws have increased the availability of information to citizens, with the aim of improving control over government actions.

However, corrupt practices in public procurement continue to be a significant problem in many countries, sometimes even surpassing past cases of corruption. In Colombia, sectors such as health, education, agriculture, infrastructure, mining, tax collection and pensions have been affected by corruption in recent years. Tackling this corruption requires collaborative citizen oversight, as government watchdogs often lack sufficient resources to monitor all contracting processes.

For citizens to be able to demand accountability, they need to have access to information about the actions of government and officials. In this regard, the effective use of new technologies to combat corruption in public procurement is proposed. Although information on public procurement processes is largely available, processing it is hampered by the quantity and velocity of data, as well as the non-machine-processable formats in which it is presented.

On the other hand, crowdsourcing, an approach that harnesses the capabilities of the crowd to solve problems, has become a useful tool in data acquisition and information generation.

This book reports results of research that focused on how to use crowdsourcing to collect data on public procurement processes from unstructured information on the web. The results and process of the research are presented in different sections, including the problem statement, the methodology, the development of a crowdsourcing application and the evaluation of the collected data.

CHAPTER 2

2. PROBLEM

To explain the relationship between governments and citizens, one can use the Agency theory, which according to Smith[1] applies to any situation where one or more persons have delegated decision-making authority to another person/s. This theory illustrates the relationship between two fundamental actors: the principal, who represents those who delegate authority, which in this case will be the citizens; and the agent who represents those who receive the delegated authority, i.e. the elected public officials. In this relationship the citizens are the principals and in contrast the governments are agents working for the citizens. By delegating some decision-making power to the agent, the principal expects the agent to defend its interests and maximise its profits by properly managing the discretionary capacity to act that has been given to it.

Agency theory allows us to analyse the problem of corruption that is facilitated when the agent possesses more information than the principal, which according to Schedler[2] is called **information asymmetry**. It is indisputable that the principal has the right to know exactly all the actions of the agent (or all the information he possesses) so that the agent does not act against his interests. But, when information asymmetries are created in the principal-agent relationship, it facilitates the corruption of the agent, who may act to favour his personal interests to the detriment of the principal's interests.

In order to put an end to the monopoly of information and seeking to reduce the discretion of the agent, many countries, through eGovernment and *Freedom of Information* (FOL) laws, have increased transparency by making public information available to citizens, including procurement information. However, a problem that currently arises when trying to **extract knowledge** about what happens with public procurement is that the information provided is intended for human reading and interpretation and in most cases it is not published in machine-processable formats, a situation that makes it difficult to process automatically with modern computer techniques.

Having open and comprehensive datasets of procurement processes in appropriate formats has become an unfulfilled promise by governments that is still far from being fulfilled. Recently, as many countries are joining the *Open Government Data* (OGD) initiative, small open data catalogues are becoming available, in some cases providing structured information, but this is the exception rather than the rule.

On the other hand, taking into account that crowdsourcing is currently used for data collection and information generation, this research focuses on whether the use of crowdsourcing as a new method of data collection is suitable for the generation of (structured) datasets in the public procurement domain.

Based on the combination of the different elements presented, the following question

---

[1] SMITH, Robert, and BERTOZZI, Mark. Principals and agents: An explanatory model for public budgeting. Journal of Public Budgeting. In: Accounting & a Financial Management. 1998, vol.10, no 3, p.325.

[2] SCHENDLER, Andreas, Plattner, Mark, and Diamond, Larrry. Conceptualizing Accountability in The self restraining state: power and accountability in new democracies. s.l : Lynne Riener Publishers, 1999. p.395.

is formulated as the backbone of the research:
How to generate datasets of public procurement processes from unstructured information published on the web using crowdsourcing as a data collection method?

CHAPTER 3

3. JUSTIFICATION

Given the relevance and magnitude of the problem of corruption in public procurement, it is necessary to work on all possible fronts to combat it, the question is how to do so. Although there are still no definitive answers to this question, when it comes to understanding and tackling something so complex, Mao TSE-TUNG's approach to contradictions can be applied[3] , which indicates that when a contradiction is complex, it is because other contradictions are involved and it is necessary to find the main contradiction, since this influences the existence and development of other contradictions. Therefore, when studying any contradiction in which other contradictions are embedded, we must devote all efforts to finding the main contradiction, because once this main contradiction is understood, all other contradictions can be resolved.

For the problem of corruption, Fan[4] states that transparency is the main contradiction, explaining that the lack of it does not allow society to know the actions of state entities and their officials, creating conditions that favour corrupt practices by making them less risky and more attractive. This is why in many countries, including Colombia, information on public procurement has been made available to citizens through e-Government, increasing transparency. However, as mentioned, it is possible that the problem does not lie in transparency as such, since the availability of information does not guarantee that the most relevant facts, including indications of anomalies and cases of corruption, will be known in order to address the problem of corruption in public procurement.

Oriol[5] points out the role that technology plays in corruption, and authors such as Criado[6] state that it is necessary to investigate how new technologies can be used effectively to help citizens fight and prevent corruption. In this sense, Web 2.0 or the collaborative Web has great potential that needs to be explored.

As previously mentioned, this research proposal investigates the use of crowdsourcing as a new method of data collection in the field of public procurement, with the aim of using this new method to generate data sets that can later be used to develop analytical models (predictive and descriptive) that help to understand and/or know what is happening in the procurement under study.

It should be pointed out that currently around the world, procurement information is not published in full, but it can be used to carry out important studies such as those carried out by Fasekas[7] , where, based on published data and with the help of a small group of collaborators, he manages to measure the *corruption risk index* (CRI).

---

[3] TSE-TUNG, M. On Contradiction [online]. August 1937 [accessed 8 February 2016]. Available online: http://www.marxists.org/reference/archive/mao/selected-works/volume- 1/mswv1_17.htm.

[4] FAN, y., ZHANG, z., and YUE, q. E-government, Transparency and Anti-corruption: How can Egovernment reduce corruption effectively in China?. In: *International* Conference on Management of e-Commerce and e-Government. s.l: IEEE, 2009. p.101.

[5] ORIOL, P, Joan. New technologies, democracy and efficiency of political institutions. In: Revista de Sociología. 2005, vol.78, p.151.

[6] CRIADO, Op. Cit. p.2037.

[7] FASEKAS, Mihály, TÓTH, János, and KING, Peter. Anatomy of grand corruption: A composite corruption risk index based on objective data. Budapest: Corruption Research Center Budapest, 2013. p.9.

Similarly, authors such as Barr[8] state that the use of models to study corruption can yield important results and impacts on society despite the lack of complete first-class data, and if one takes into account that corrupt practices leave their mark on official records (according to Fasekas' experience[9] in his study of corruption), then there are currently some basic (not ideal) conditions that allow studies such as the one proposed in this research to be carried out.

[8] BARR, Joshua, PINILLA, Edgar, and FINKE, Jorge. A Legal Perspective on the Use of Models in the Fight Against Corruption, In: South Carolina Journal of International Law and Business. 2012, vol.8, no.2, p.266.

[9] FASEKAS, Mihály, TÓTH, János, and KING, Peter: Corruption manual for beginners: Corruption techniques in public procurement with examples from Hungary. Budapest: Corruption Research Center Budapest, 2013. p.10.

CHAPTER 4

4. OBJECTIVES

4.1 GENERAL OBJECTIVE

To develop a crowdsourcing-based data collection method that allows the generation of datasets of public procurement processes from unstructured information published on the web.

4.2 SPECIFIC OBJECTIVES

- Define the stages, elements and characteristics of the proposed crowdsourcing-based data collection method to generate datasets of public procurement processes.
- Develop a crowdsourcing application prototype that supports the defined method and allows volunteers to participate in the transcription of public procurement data by generating raw data from the indicated procurement processes.
- Design and implement a procedure and techniques to obtain final data resulting from the aggregation and consensus of the raw data generated by individuals for public procurement processes.
- Evaluate the proposed method and the accuracy of the final data against reliable data indicating whether this new data collection method is suitable for the generation of datasets in the public procurement domain.

CHAPTER 5

5. REFERENCE FRAMEWORK

5.1 CONCEPTUAL FRAMEWORK

The project takes place in the context of e-Government, which makes it necessary to start with a precise definition of this term. From there, other key ideas offered by e-Government will be explored, such as transparency, corruption and accountability. It will also introduce the concept of Web 2.0 and crowdsourcing, which are areas of focus for this project. In addition, some additional concepts related to mid-chain and evolutionary algorithms, which emerged during the development of the project and were considered relevant to its understanding, will be addressed.

5.1.1 E-Government.

In the Ibero-American sphere, e-Government is defined as electronic administration, as established by the Latin American Centre of Administration for Development (CLAD), which describes it as "the use of ICTs in government bodies to improve the information and services offered to citizens, to guide the effectiveness and efficiency of public management and to substantially increase public sector transparency and citizen participation"[10] . Mahmood[11] stresses that e-Government improves the interaction between governments and citizens, while Ndou[12] identifies a number of associated opportunities, such as cost reduction, increased efficiency, improved quality of services, transparency, fighting corruption, accountability, strengthening government capacity, building networks and communities, improving decision-making and promoting the use of ICTs in a variety of ways in society.

5.1.2 Transparency

For Relly and Sabharwal[13] in the literature there is no unanimity on the definition of transparency, but it can be defined according to Florini[14] as the degree to which information is made available to people outside the government administration so that people can influence decisions. The information released by transparency is intended to allow many eyes to scrutinise government entities. This is why Valcerde[15] indicates that the concept of transparency involves the opening and flow of information from public bodies into the public domain, allowing potential stakeholders to review and analyse the information in order to detect possible anomalies.

Tesoro[16] also states that transparency is the duty of the political powers to submit

---

[10] CLAD. Ibero-American Charter on e-Government. [online]. Santiago de Chile: November 2007 [accessed 8 February 2016]. Available online: http://old.clad.org/documentos/declaraciones/cartagobelec.pdf/view

[11] MAHMOOD, Rumel. Can information and communication technology help reduce corruption? How so and why not: Two case Studies from South Asia. In: Perspectives on Global Development and Technology, 2006, vol.3, no.3, p.347.

[12] NDOU, Valentina. E-Government for Developing Countries: Opportunities and Challenges. In: The Electronic Journal of Information Systems in Developing Countries (EJISDC), 2005, vol.18, no.1, p.15.

[13] RELLY, J. E., and SABHARWAL, Meghna. Perceptions of transparency of government policymaking: A cross-national study. Government Information Quarterly, 2009, vol.26, no.1, p.148.

[14] FLORINI, Ann. The Right to Know: Transparency for an Open World. s.l: Columbia University Press, 2007, 376 pp.

[15] VALCERDE, M. (2004). Transparency, access to information and accountability: Experiences from the European Union and Mexico. In: IX Congreso Internacional del CLAD sobre la reforma del estado y la administración pública. Madrid: Clad, 2004, p.8.

[16] TESORO, José. Transparencia y tecnologías de gestión: la viabilidad de sistemas con transparencia en la administración pública. In: Estado y Sociedad, vol1, no.33, p.33.

information about their management, the management of public money entrusted to them by society and the criteria they use to support their decisions to the public for analysis. Transparency should make it possible to know how the nation's budget is formulated, executed and evaluated, because it is all taxpayers' money and it is this information that requires total transparency. This is why Gascó[17] says that transparent governments do not hide information and are open to public scrutiny.

5.1.3 Accountability

Transparency plays a crucial role in promoting accountability, understood as the ability of citizens to scrutinise and evaluate the actions of government entities and their representatives. According to Cees and Mark[18] , political accountability implies that politicians are responsible for their actions and are rewarded or punished accordingly.

5.1.4 Corruption

Transparency International defines corruption as the misuse of power for personal gain, a definition that, although general, sheds some light on the concept. It is important to note that corruption refers to the improper handling of public funds and fraud in the area of private funds.

On the other hand, the United Nations[19] describes corruption as the result of a combination of monopoly of power and unaccountable and discretionary decision-making. This definition highlights two key aspects that are conducive to corruption: concentration of power and lack of transparency. Other general definitions, such as "the abuse of public power for private gain"[20] , fail to fully capture the concept due to its complexity, as discussed at the International Criminal Law Congress of the International Association of Penal Law, as the notion of corruption can vary considerably from country to country.

The World Bank[21] interprets the term abuse as accepting or soliciting bribes for personal gain regardless of public benefit, noting also that corruption can be divided into: political and bureaucratic corruption, isolated and systemic corruption, and private sector corrupt practices. For Nicholls et al.[22] political and bureaucratic corruption is corruption that involves the misappropriation of funds from the wealth of the state, carried out by politicians or public officials (also called grand corruption).

De Graaf and Huberts[23] indicate that isolated corruption applies to situations of corruption that occur individually and not as a norm in society, and systemic

---

[17] GASCÓ, Mila. The role of institutions in the development of e-government in Latin America: some reflections. In: Revista Chilena de Administración Pública, 2009, vol.1, no.14, p.37.

[18] CEES, Van der Eijk and Mark Franklin. Elections and voters: Assessing electoral democracy. London: Palgrave Macmillan, 2009, p.213.

[19] UNITED NATIONS. United Nations Convention Against Corruption. [online] New York: 2004. [accessed 09 February 2016]. Available online: http://www.unodc.org/pdf/corruption/publicationsunodcconvention-s.pdf

[20] BUSCAGLIA, Edgardo, and DIJK, Van. Controlling Organized Crime and Corruption in the Public Sector. In: Forum on Crime and Society, 2004, vol.3, no.1, p.3.

[21] WORLD BANK. Helping Countries Combat Corruption: The Role of the World Bank. [online]. September 1997. [accessed 9 February 2016]. Available online: http://www1.worldbank.org/publicsector/anticorrupt/corruptn/cor02.htm

[22] NICHOLLS, Collins, et al. Corruption and misuse of public office. Oxford: Oxford University Press, 2006, p.20.

[23] DE GRAFF, Gjalt, and HUBERTS, Leo. Portraying the nature of corruption using an explorative case study design. In: Public administration review, 2008, vol.68, no.4, p.640.

corruption is when corrupt activities are a norm in society and occur at all levels.
Beyond the existence of precise definitions for the term corruption, what is crucial lies in its scope and repercussions, as it is a problem of great complexity and difficulty to address, given its interrelationship with various social, economic, legal and cultural problems, among others. Studies such as Oriol's[24] have highlighted that corruption is also influenced by technology-induced efficiency.
It is in this context that the development of the project focuses, specifically on the use of technologies and systems that are part of the Social Web. Concepts such as crowdsourcing are of particular interest and are described below.

5.1.5 Social Web

According to Murugesan[25] there is no universally accepted definition in the literature for the term Web 2.0 which was coined by Tim O'Reilly in 2005 and defined as follows:
Web 2.0 is defined as a platform where the network encompasses all connected devices. Web 2.0 applications take advantage of the inherent benefits of this platform by offering software as a service that is constantly being updated and enhanced as the number of users increases. These applications also allow the combination and consumption of data from a variety of sources, including individual users, and provide their own data and services so that they can be combined by others. This approach encourages networking through a participatory architecture and goes beyond the Web 1.0 metaphor of the page, seeking to provide enriching experiences for users .[26]
Porter[27] examines the above definition and points out that in Web 2.0 (also known as the Collaborative Web or Social Web), traditional desktop applications are abandoned in favour of new services offered through web platforms and technologies that are not tightly integrated, but allow user interaction. This contrasts with conventional media that have a single address and/or limited back-and-forth interaction with the consumer. On the other hand, social media enable diverse and extensive many-to-many interaction, i.e. between users with each other and between users and entities. In the Social Web, the term "social media" is used to refer to media generated through the social interaction of users via Web 2.0 technologies (encompassing user-created content). The most recognised social media include blogs, wikis, forums, social bookmarking, collaborative editing tools and social networking services.

5.1.6 Crowdsourcing

The term crowdsourcing was first coined in 2006 in Wired magazine by Jeff Howe, who described it as "the action of outsourcing a function normally performed by an employee of a company or institution to an undefined (and usually large) group of people through an open call"[28] . Brabham[29] , for his part, describes crowdsourcing as

---

[24] ORIOL, Op. Cit. p.173.
[25] MURUGESAN, San. Understanding Web 2.0. In: IT Professional, 2007, vol.9, no.4, p.34.
[26] O'REILLY, Tim. Web 2.0: Compact Definition? [online] October 2005. [accessed 9 February 2016]. Available online: http://radar.oreilly.com/2005/10/web-20-compact- definition.html
[27] PORTER, Joshua. Designing for the Social Web. s.l: New Riders Press, 2008, p.5.
[28] HOWE, Jeff. The rise of crowdsourcing. In: Wired Magazing, 2006, vol.14, no.6, p.61.
[29] BRABHAM, Darem. Crowdsourcing as a model for problem solving: An introduction and cases. In: Convergence: The International Journal of Research into New Media Technologies, 2008, vol.14, p.75.

a new business model especially useful for tasks that require a wide variety of different viewpoints or solutions to problems, and which is generally carried out online.

Since the term emerged, several definitions have been proposed to represent its evolution. Finding that its conceptual meaning can cause confusion with other nearby terms, such as open innovation, also co-creation and the most controversial: collective intelligence. Some relate it to human or social computing, indicating that it is an emerging paradigm in computing. In order to provide conceptual clarity, Estellés and González proposed an integrative definition of crowdsourcing after analysing more than 40 published definitions.

Crowdsourcing is defined as a participatory online activity in which a person, institution, non-profit organisation or company invites a group of individuals, of varying knowledge, skills and quantity, through an open and flexible call, to voluntarily perform a specific task. This task may vary in its complexity or modularity, and requires the participation of the crowd, who may contribute labour, money, knowledge and/or expertise. This collaboration always implies a mutual benefit: the user obtains the satisfaction of a specific need, be it economic, social recognition, self-esteem or personal development, while the crowdsourcer receives and uses the user's contribution for his or her benefit, the form of which will depend on the type of activity. [30]

Estellés and González[31] also propose 8 criteria to assess whether an initiative is crowdsourcing or not: (1) The crowd (people who participate in the tasks); (2) crowdsourcer, entity (person or organisation, etc.) that postulates the tasks; (3) The task, what is required to be done; (4) The platform, software system (usually) used to support the process.

Crowdsourcing is currently in its infancy and is evolving as new applications emerge.

5.1.7 Medium Chain

The initial definition of mean chain[32] was presented by Kohonen[33] who first defined the concept as follows: given a set of chains, the mean chain is defined as a chain that has the sum of the minimum distances of all the chains in the set. The above definition is represented by the following mathematical expression (equation 1):

$$m_S = \underset{s \in \Sigma^*}{\operatorname{argmin}} \sum_{i=1}^{n} d(s, s^i) \qquad (1)$$

Where ms denotes the mean string to be found given a set of strings S1= {S^1 , S2, ... , S^n } consisting of elements of an alphabet Σ, where Σ* denotes that the strings have finite length and d is the dissimilarity or distance measure defined according to the application.

Figure 6-1 is intended to illustrate that the average chain to be found from a set of given chains must satisfy the condition that the sum of the distances from it to all the

[30] ESTELLÉS, Enrique, GONZÁLEZ, LADRÓN-DE-GUEVARA. Towards an integrated crowdsourcing definition. In: Journal of Information Science, 2012, vol.38, no.2, p.189.

[31] Ibid., p. 190.

[32] In this work, the concept of median string is expressed by the generalised term median string, which can lead to confusion because the translation of median is median, but the concept actually refers to the middle string.

[33] KOHONEN, Teuvo. Median strings. *Pattern Recognition Letters*, 1985, vol. 3, no 5, p. 309-313.

given chains must be minimum.

Figure 5-1. Representation of distances to the middle chain

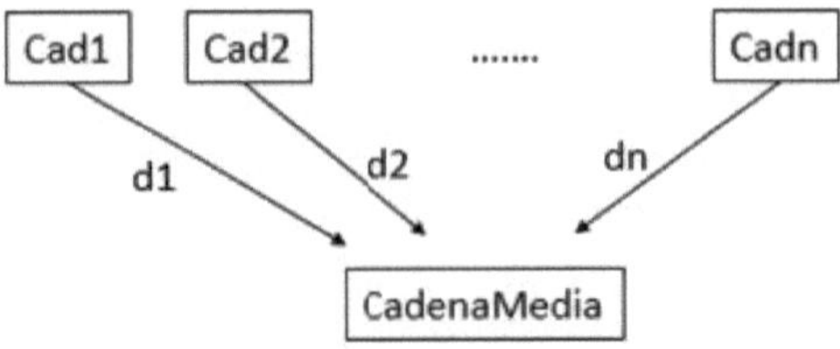

Source: own elaboration

The d¿ distances can be measured in several ways but the best known is the Levenshtein metric, which allows finding the distance between two strings, i.e. it is a measure that represents how dissimilar the strings are. Table 6-1 gives examples showing that as the Levenshtein distance increases, the words become less similar.

Table 5-1. Example of chain spacing

Chain 1	Chain 2	Distance
house	house	0
house	hunting	1
cease	hunting	2
house	**dog**	5
abed	**should**	4

Source: own elaboration

When trying to find the average string, a search space must be generated and explored, consisting of all the strings that can be created with all the possible combinations of the elements of the alphabet Σ. The problem then is to search this space until a string is found whose sum of distances with respect to the set of input strings is the minimum.

Evolutionary algorithms can be used to tackle the solution of problems such as finding the average chain, which is precisely what is discussed below.

5.1.8 Evolutionary Algorithms

Evolutionary algorithms are historically considered a branch of artificial intelligence since they were used in the 1960s to add intelligence to finite state machines, but today they are used to solve a wide variety of problems that are computationally difficult to solve, i.e. those that do not have a solution in polynomial time, such as for example: the travelling agent problem, the knapsack problem, the problem of finding a Hamiltonian cycle, and real-life problems such as: optimisation of rod cutting, optimisation of fabric cutting, optimisation of car park areas, etc. All these problems for relatively small inputs the response time executed on the best computer in the world would take the time in which the universe was created until today. Therefore it is necessary to attack these problems with alternative strategies where the response time is reasonable.

Listed below are some different approaches or ways to implement evolutionary algorithms:

- Genetic algorithm: these are the algorithms created by Holland, which initiated evolutionary computation. It is characterised by the fact that the chromosome is encoded in zeros and ones, and all functions are applied to expressions of zeros and ones. You must have a decoding function.
- Genetic programming: used to create programs automatically to solve specific problems, the whole structure is based on trees.
- Grammatical evolution: from some grammatical rules, programmes are created by generating other, more robust grammatical rules and tested for the specific problem.
- Gene expression programming: a program is created with the concept of a tree but its expression is a vector of characteristics from that tree read from left to right by level.
- Evolutionary programming: very similar to the genetic algorithm but the chromosome is not coded, the phenotypic part is the same as the genotypic part. Operators are defined on a case-by-case basis if applicable. The original approach was applied in finite state machines.
- Evolutionary strategies: there is an objective function with many parameters of floating type, these parameters are slowly modified with a strategy called Gaussian noise. The survival of individuals is based on keeping the parameters of those that best meet the objective.
- Classifier systems: it consists of creating new rules from a few rules through the genetic algorithm, which is the one that makes the combinations and the aptitude tests of the new rules created. It is similar to the learning of an expert and in the end it becomes an expert system.
- Simulated cooling: each individual is accepted if it fulfils at least one of the following conditions: if it is better than the previous one it is accepted and if it is worse then it is accepted as long as it fulfils a probability cooling function, i.e. even if the individual is bad initially it can be accepted, but over time if it does not improve then it will be rejected because the acceptance threshold becomes smaller and smaller.

The approach selected for this project was evolutionary programming because a character string will be used as the solution representation and the inputs are also character strings, which implies that there is no encoding or decoding function for the strings being worked with.

5.2 THEORETICAL FRAMEWORK

This is followed by other theories that highlight the possibilities offered by the Internet and certain technologies linked to the Social Web in the fight against corruption. The use of crowdsourcing in data collection, the problems related to the calculation of the average chain, and a brief description of the general functioning of a genetic algorithm are also discussed.

5.2.1 Social web technologies and their use in the fight against corruption Shim and Eom[34] argue that Web 2.0 social media are considered to be one of the most

---

[34] SHIM, Dong, EON, Tae, HONDEGHEM, A. and PERRY, James. Anticorruption effects of information communication and technology (ICT) and social capital. In: International Review of Administrative Sciences, 2009, vol.75, no.1, p.99.

effective tools in the fight against corruption. According to Bertot et al.[35] , these media enable the creation of social capital, which is fundamental to reducing corruption by fostering a culture of transparency and detection of corrupt practices. In this context, Chen[36] points out that the use of these technologies allows citizens to overcome the restrictions imposed by the elites that control the traditional media, facilitating the right to free expression of any individual interested in public affairs who wishes to expose criticisms, facts and figures related to possible acts of corruption.

The Social Web acts as a megaphone that amplifies public attention to these cases, while providing anonymity to participants and enabling the rapid dissemination of information on corruption in digital format, overcoming the limitations of time and space. Moreover, the interactivity of social media facilitates the attraction and mobilisation of large numbers of people, leveraging the power of crowds, who can easily follow, comment and gather evidence of corruption through platforms such as forums or blogs.

5.2.2 Quality of Crowdsourcing results

According to Erickson[37] , crowdsourcing represents a novel approach to tackling problems by harnessing the perceptual and cognitive skills of a wide group of people through Web 2.0. The Internet has played a crucial role in the development and evolution of this concept by providing the technology that supports Web 2.0. This allows such initiatives to have access to a large number of individuals from all over the world, which considerably broadens their scope and diversity of participants.

By applying the concept of crowdsourcing, a wide variety of applications have been developed on more than 70 existing platforms in recent years, such as Amazon Mechanical Turk, Crowdflower and CloudCrowd. A common practice in the use of crowdsourcing involves posting tasks as queries to crowd workers and offering them a reward (such as financial incentives or recognition).

According to Quinn[38] , a major challenge for crowdsourcing applications is the quality control of the results obtained. Given the open nature of crowdsourcing, the data collected through this process may contain errors. Hence, several strategies are available to contribute to quality, such as:

(1) Redundancy / Aggregation of results: According to Duan[39] , the strategy of aggregating results involves introducing redundancy by asking several people to carry out the same work and then pooling the results to produce a more reliable answer. In short, the aim of aggregation is to uncover the hidden truth within the responses provided by workers to the question or activity.

(2) Golden data: According to Oleson[40] , quality control using golden data is based

---

[35] BERLOT, op. cit. p 266.

[36] CHEN, op. cit. p 87.

[37] ERICKSON, Thomas. Geocentric Crowdsourcing and Smarter Cities: Enabling Urban Intelligencein Cities and Regions. In: 1st International conference on ubiquitous computing. Copenhagen: ACM, 2010, p.4.

[38] QUINN, Alexander, and BEDERSON, Benjamin. Human computation: a survey and taxonomy of a growing field. In: Conference on Human Factors in Computing Systems. Vancouver: ACM, 2011, p.1403.

[39] LEI, Duan, SATOSHI, Oyama, HARUNHIKO, Sato, and MASAHITO, Kurihara. Separate or joint? Estimation of multiple labels from crowdsourced annotations. In: Expert Systems with Applications, 2014, vol.41, no.13, p.5723.

[40] OLESON, David, et al. Programmatic gold: Targeted and scalable quality assurance in

on so-called golden questions, the answers to which are previously known, and which can be automatically evaluated to assess the quality of the work performed. This validation process is simple and requires no additional time or human intervention. The percentage of correct answers is checked against a threshold set by the crowdsourcer applicant to determine the acceptance or rejection of the answers provided by the participant.

(3) Multi-level reviews: also called multi-level reviews according to Quinn[41] , these involve the participation of a second group of workers to corroborate the answers provided by the first group (this method can involve more than two levels).

(4) Expert review: According to Benatallah[42] , the results delivered by workers are reviewed prior to payment and acceptance. This is effective, but it does generate more costs, which affects scalability, not to mention the need for subject matter experts.

(5) Reputation and pre-selection: Paolacci[43] suggests that workers for a specific task can be pre-selected based on their reputation. They can also be asked to answer pre-defined questions to pre-test them before they perform the tasks.

(6) Other methods: For example, Tweddle[44] and colleagues emphasise the importance of exploring other methods, such as training participants. In addition, providing different resources that help and contribute to the completion of the tasks that are assigned.

Various combinations of the quality control mechanisms mentioned above can be explored to improve results. However, response aggregation emerges as the main approach to implement quality control, as it leverages the collective intelligence that can emerge from the contribution of the crowd. Despite these challenges, crowdsourcing has proven its effectiveness by harnessing the capabilities of human beings. For this reason, it is referred to as Human Intelligence Tasks (HITs).

Finally, according to Papadopoulou[45] crowdsourcing is currently being used to collect data or generate information. Proof of this is the number of crowdsourcing applications that have addressed data acquisition, data integration, data mining, information extraction, and information retrieval; as can be found in Amsterdamer[46] , Zhang[47] , Sun[48] , Gokhale[49] and Yan[50] among others.

---

crowdsourcing. In: 25th AAAI Conference on Artificial Intelligence. San Francisco: Association for the Advancement of Artificial Intelligence, 2011, p.43.

[41] QUINN, op. cit. p 1404.

[42] BENATALLAH, A., et al. Quality Control in Crowdsourcing Systems: Issues and Directions. In: IEEE Internet Computing, 2013, vol.17, no.2, p.76.

[43] PAOLACCI, Gabriele, CHANDLER, Jesse, and IPEIROTIS, Panagiotis. Running Experiments on Amazon Mechanical Turk. In: Judgment and Decision Making, 2010, vol.5, no.5, p.411.

[44] TWEDDLE, J., L. ROBINSON, D., POCOCk, M.J.O., ROY, H.E. Guide to citizen science: developing, implementing and evaluating citizen science to study biodiversity and the environment in the UK. [online] S.l: Natural History Museum and NERC Centre for Ecology & Hydrology for UK-EOF, 2012, 36 p. [accessed 9 February 2016]. Available online: http://www.nhm.ac.uk/content/dam/nhmwww/take-part/Citizenscience/citizen-science-guide.pdf

[45] PAPADOPOULOU, op. cit. p 115.

[46] AMSTERDAMER, Yael, GROSSMAN, Yael, MILO, Tova, and SENELLART, Pierre. Crowd mining. In: International Conference on Management of Data. s.l: ACM, 2013, p.241.

[47] ZHANG, Chen, CHEN, Lei, JAGADISH, H.V., and CAO, Chen. Reducing uncertainty of schema matching via crowdsourcing. In: Very Large Database Endowment. 2013, vol.6, no.9, p.757.

[48] SUN, Chong, RAMPALLI, Narasimhan, YANG, Frank, and DOAN, Anhai. Chimera: Large-scale classification using machine learning, rules, and crowdsourcing. In: Very Large Database Endowment.

According to Papadopoulou[51] , crowdsourcing is currently being used in data collection and information generation processes, representing a relatively novel approach to the knowledge generation process. This is evidenced by the number of crowdsourcing applications that have focused on data and information collection, integration, mining, extraction and retrieval, as observed in works by Amsterdamer[52] , and Zhang[53] , among others.

5.2.3 Calculation of the Mean Chain

The concept of mean string is equivalent to that of mean vector but in this case the dataset is made up of character strings, so its calculation is not simple. According to Nicolas[54] the calculation of the average string is of great significance in several areas of research because in other words, the average string summarises the information shared by the initial set of strings from which it was calculated and depending on the application, it subsequently serves for example as: an index in searches in the case of databases and data mining, as a pattern that is used to search in longer texts (computational biology), or as described by Gusfield[55] for classification purposes in speech recognition and computational biology.

In this work, the average string is used to represent the aggregation of text entries generated by different people in Crowdsourcing initiatives. As finding the exact average string may not be a feasible option, when tackling this problem, solutions that allow finding a good approximation but with a reasonable computational cost must be worked out.

5.2.4 General operation of an evolutionary algorithm

The general operation of an evolutionary algorithm consists of the following steps:

- **Initialisation**: a set of chromosomes with random values is created, where each chromosome is a possible solution to the problem to be solved or optimised. The set of chromosomes generated in the initialisation are stored in a memory space called a warehouse.
- **Evaluation:** in this phase, a fitness value or rating is given to the chromosome to see how close it is to solving the problem.
- **Selection for breeding**: a subset of the existing population of chromosomes in the cellar is selected and then subjected to a series of functions generally referred to as breeding (crossover or mutation) or survival functions[56] . There are several

---

2014, vol.7, no.13, p.1529.

[49] GOKHALE, Chaitanya, et al. Corleone: Hands-off crowdsourcing for entity matching. In: ACM SIGMOD International Conference on Management of Data. Snowbird: ACM, 2014, p.601.

[50] YAN, Tingxin, KUMAR, Vikas, and GANESAN, Deepak. Crowdsearch: Exploiting crowds for accurate real-time image search on mobile phones. In: 8th international conference on Mobile systems, applications, and services. San Francisco: ACM, 2010, p.77.

[51] PAPADOPOULOU, op. cit. p 115.

[52] PAPADOPOULOU, op. cit. p 115.

[53] SUN, Chong, RAMPALLI, Narasimhan, YANG, Frank, and DOAN, Anhai. Chimera: Large-scale classification using machine learning, rules, and crowdsourcing. In: Very Large Database Endowment. 2014, vol.7, no.13, p.1529.

[54] NICOLAS, François; RIVALS, Eric. Complexities of the centre and median string problems. In: Annual Symposium on Combinatorial Pattern Matching. Springer Berlin Heidelberg, 2003. p. 315.

[55] GUSFIELD, Dan. Algorithms on strings, trees and sequences: computer science and computational biology. Cambridge university press, 1997.

[56] The term survival function comes from biology when dealing with the permanence or extinction of

strategies for selecting chromosomes after establishing how many or what percentage will be passed for breeding, for example:

- Choose the ten best chromosomes, i.e. the ten with the best fitness function.
- Randomly select 10 chromosomes from the chromosome pool.
- Sampling by lot, where the chromosomes with the best fitness function are more likely to be chosen.
- Using the roulette method, a random value is generated and from this value, through a simple recurrence relation, other missing random values are formed.
- By tournament, selecting three chromosomes at random and from these the one with the best fitness value is chosen and the procedure is repeated until the set number of chromosomes is chosen.

- **Recombination**: different reproductive functions are applied to selected chromosomes in order to generate new offspring. Among the functions to be applied are crossover and mutation. Not all the descendant chromosomes will be good, but the recombination process allows some better chromosomes to be generated. This procedure is done until a point is reached where the new chromosomes are no better than the ones in the cellar, a signal that serves to stop the evolutionary algorithm.
- **Replacement**: there are many ways to replace old chromosomes in the cellar with new descendant chromosomes, the most commonly used of which include:

- The ten chromosomes generated will replace the ten baddest chromosomes in the cellar.
- If many chromosomes are generated then the whole cellar is replaced with all the descendant chromosomes.
- Only chromosomes that outperform or beat the worst chromosomes are replaced; many of the chromosomes generated are worse than the worst chromosomes in the cellar.

When implementing an evolutionary algorithm, the elements shown in Figure 6-2 must be generated.

Figure 5-2. Main elements to be implemented in evolutionary computation

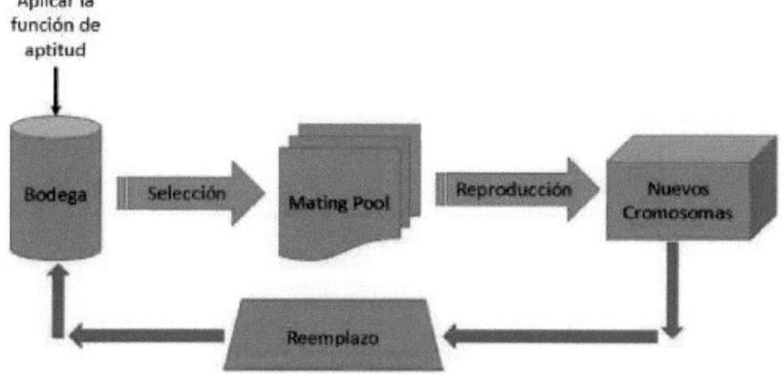

Source: own elaboration

These elements appear during the execution of the evolutionary algorithm as follows:

---

individuals.

the pool is created in the initialisation containing the chromosomes to which the fitness function is applied in the evaluation. Subsequently, during selection, the chromosomes that are selected are copied to the mating pool. To proceed with the reproduction where new chromosomes are created that during the replacement can take the place of the worst chromosomes in the storage, completing the process that can be repeated until the new chromosomes do not improve the ones in the storage or a defined number of iterations is completed.

As there are n possible solutions in the warehouse, all algorithms to be implemented must be computationally low cost and also effective.

5.3 LEGAL FRAMEWORK

5.3.1 Transparency and Free Access to Public Information in Colombia

Laws and regulations on transparency and free access to public information are fundamental to promoting accountability, strengthening democracy and ensuring open and accountable government both in Colombia and internationally.

5.4 STATE OF THE ART

5.4.1 Current status of public procurement data available on the web

Internet and related technologies have reduced the cost of collecting, distributing and accessing government information including procurement information.

Information and records on public spending are usually released through dedicated portals or websites known according to Matheus[57] as transparency portals; these portals are related to the e-Government sectors *Government-to-Citizen* (G2C) and *Government-to-Business* (G2B).

Currently, many countries have transparency portals where procurement information is published, some of them are part of e-procurement systems and there are two types: newer ones that support the entire procurement process and older ones that are basically a repository for documents generated in procurement processes that are carried out physically.

The current situation with regard to procurement data published by eGovernment is similar in many countries around the world, where data is published and made available to citizens in formats that are not machine-processable, because they are intended for human reading and interpretation, a condition that makes automatic processing with modern computational techniques difficult.

Recently, as many countries are joining the *Open Government Data* (OGD) initiative, some open data sets in structured formats are becoming available for public procurement, but these do not necessarily include or cover all information associated with procurement processes.

The format is not the only problem with public procurement data; another problem is the lack of standardisation in aspects such as the information to be published. For this reason, the World Bank, seeking to solve different problems that currently arise with the information that the e-Government handles on public procurement, supported the *Open Contracting Partnership* (OCP)[58] in the creation of the *Open*

---

[57] MATHEUS, Ricardo, RIBERIRO, Manuella, VAZ, José, and SOUZA, Cesar. Internet Use for Social Control and Participation: What Are Local Governments Doing in Latin America? In: 4th International Conference on Theory and Practice of Electronic Governance. Beijing: ACM, 2010, p.333.

[58] http://www.open-contracting.org/

Contracting Data Standard (OCDS)[59] which is an open data standard for the publication of structured information covering all stages of a procurement process: from planning to execution.

Some countries are currently working towards implementing best practices in the management of public procurement information and are adopting OCDS, and it is expected that new generations of e-procurement systems will support it so that stakeholders can have access to complete data in appropriate formats.

5.4.2 Use of crowdsourcing in the collection of public procurement data

In the review of the scientific literature, no studies were found in which crowdsourcing is used as a method for collecting data on public procurement processes. However, there are some projects developed in the social sphere that are interested in procurement information, but the results of these initiatives have not been published in the scientific literature. One of the most important of these is the OpenSpending initiative[60] , which seeks to facilitate the monitoring and analysis of public financial information worldwide by providing a platform for uploading and analysing information.

Crowdsourcing is a concept that is widely spread. The literature survey showed that crowdsourcing is being used in a multidisciplinary way, spanning many fields of study such as business and management, medicine, science and environmental sociology, and not limited to computer science.

Different mechanisms exist to control the quality of the results, see section 5.2.2.

[59] http://standard.open-contracting.org/latest/en/#

[60] https://openspending.org/

CHAPTER 6

6. METHODOLOGY

In order to achieve the proposed objective, the following methodology is proposed.

Table 6-1. Proposed research methodology

No	STAGE NAME	PURPOSE
1	Definition of the proposed method	Define the stages, elements and characteristics of the proposed crowdsourcing-based data collection method to generate datasets of public procurement processes; the results of this stage are presented in section 7.
2	Development of the crowdsourcing prototype application	This stage aims to develop a small crowdsourcing application prototype that allows testing some important elements that are identified once the proposed method is available ; the results for this stage are presented in section 8.
3	Generation of raw data	The developed prototype application is used and with volunteers some raw data is generated for the selected public procurement processes; the results for this stage are presented in Annex D and discussed in section 10.
4	Development of the quality control mechanism	Design and implement a mechanism to control quality based on aggregating results; the results for this step are presented in section 9.
5	Final data generation	Test the quality control mechanism defined in the previous step with some data.
		The results for this step are presented in Annex E and discussed in section 10.
6	Evaluation of the proposed method and the accuracy of the final data	Evaluate the proposed method and final data against reliable data (generated for this purpose are provided in Annex C); the results for this step are presented in section 10.

CHAPTER 7

7. PROPOSED DATA COLLECTION METHOD

This section presents the crowdsourcing-based data collection method that was designed to generate datasets of public procurement processes from unstructured information published on the web.

7.1 Process defined for the design of the method

The first step when starting with the design of the method was to define a procedure that would facilitate its elaboration. The steps followed were as follows:

- Describe how data collection tasks are performed on widespread Crowdsourcing platforms such as Amazon Mechanical Turk[61] (MTurk) and Crowdflower .[62]
- Describe the characteristics of Crowdsourcing based data collection methods that are found by conducting a review of the scientific literature.
- Describe the characteristics of different transparency portals (in Colombia and other countries) in relation to the publication of public procurement data.
- Describe how the manual data collection would be carried out under the conditions and characteristics presented by the transparency portals.
- Define the stages, elements and characteristics of the crowdsourcing-based data collection method to be designed.

In carrying out the above activities, a number of interesting results were found, as well as some surprises:

- In the case of Crowdsourcing platforms, it was found that MTurk is a platform that could be called general purpose because it works as a marketplace that connects companies with individuals to perform all kinds of tasks, including tasks that computers are currently unable to do. Crowdflower is a platform that specialises in data processing, integrating people and machines in these processes. The Crowdflower platform allowed us to identify how to approach crowdsourcing tasks and even how to implement a reputation-based quality control mechanism (this important aspect is discussed in section 12).
- When reviewing the scientific literature, no Crowdsourcing-based data collection methods were found that could serve as a general reference, although there are numerous works that employ crowdsourcing for different tasks related to data collection.
- When reviewing different transparency portals where contracting information is published, it was identified that these are sometimes part of the electronic contracting systems, finding two types: the newer ones that support the entire contracting process and other older ones that are basically a repository for the documents generated in the contracting processes that are carried out in a physical form. On the other hand, it was identified that these portals have forms that allow the selection of some criteria to guide the search for procurement information and the results of the queries are paginated with some minimum information of all the processes found, giving the option of accessing through links to extended information of the process that can be found directly on the page or in documents that can be downloaded,

[61] https://www.mturk.com/mturk/welcome
[62] https://www.crowdflower.com/

generally in pdf format. Finally, a very important feature is that some portals incorporate the CAPTCHA (*Completely Automated Public Turing test to tell Computers and Humans Apart*) mechanism to prevent the automatic download of information from being programmed. Figure 8-1 and Figure 8-2 show the CAPTCHA validation requested by the electronic public procurement system (Secop) in Colombia when trying to access procurement information.

• The process of manual collection of procurement data under the conditions and characteristics presented by the transparency portals is quite simple for a human, where basically first the interested party must access the URL of the transparency portal and in the search form must indicate some criteria that allow selecting the procurement processes of interest, once the query is made, different data of the processes found are available and can be copied.

- Based on the information obtained, we proceeded to define the stages, elements and characteristics of the crowdsourcing-based data collection method described below.

Figure 7-1. CAPTCHA validation requested by Secop to access contracting data

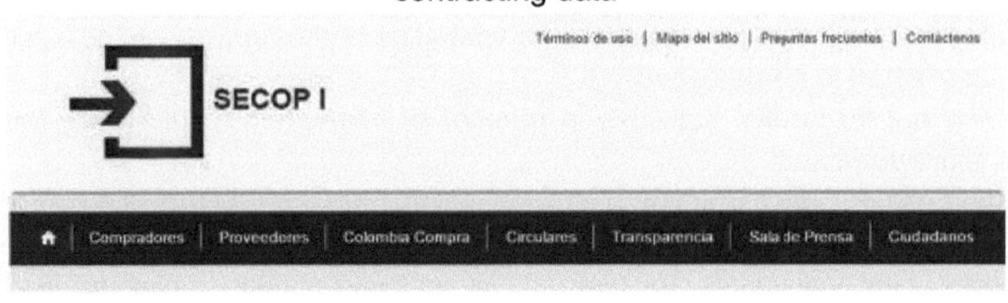

Por favor ingrese la validación de CAPTCHA para la consulta del detalle del proceso

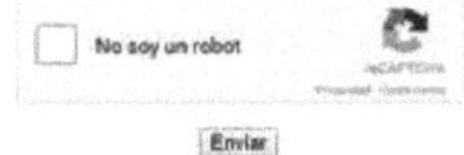

Enviar

Source: Taken from Secop

Figure 7-2. Example of CAPTCHA validation used by Secop

Source: Taken from Secop

7.2 Description of the proposed method

The method described below is suitable for use in countries[63] where governments make public procurement data available to citizens in non-machine-processable formats.

Using a "*top-down*" approach, the proposed method is described below.

7.2.1 Overview of the proposed method

Figure 7-3. shows the stages that make up the proposed method.

[63] Among the countries in this situation are Colombia...

Figure 7-3. Stages of the proposed method

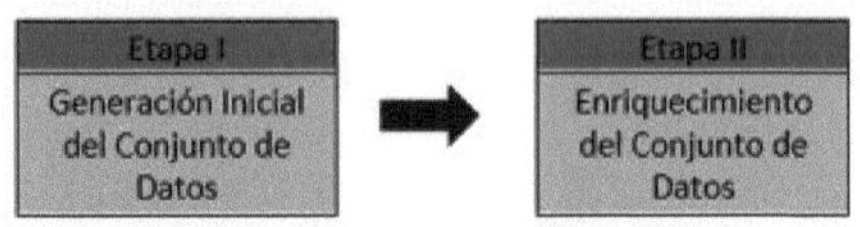

Source: Own elaboration

The first step called "initial dataset generation" is intended to allow stakeholders to generate an initial dataset containing basic data on the procurement processes of interest.

Subsequently, in the second stage called "dataset enrichment" the purpose is to complement the initial dataset by adding more intermediate level data from the processes of interest selected in the previous stage.

The data to be collected corresponds as far as possible[64] to data defined by the *Open Contracting Data Standard* (OCDS) where three levels of information disclosure are established for basic, intermediate and advanced contracting processes.

The basic and intermediate levels of the OCDS are briefly described below.

The basic level covers data that should at a minimum be handled by e-procurement systems and should be available on most transparency portals. These data provide users with an overview of procurement processes and according to OCDS are:

- Name of buyer and address.
- Procurement process identifiers, titles, descriptions, status, method of acquisition.
- Period, description of the object to be contracted and quantities.
- Documents associated with the procurement process.
- Award identifiers, titles, descriptions, dates, values.
- Names of the chosen supplier and addresses.
- Contract identifiers, titles, status, period, values, description of contractual object and quantities.
- At a basic level, the publication of contract documentation and implementation information is also recommended.

The intermediate level of the OCDS, in addition to the basic fields, has the following data:

- Organisational identifiers for buyers and suppliers
- Detailed classification of the object to be contracted, award and contracts.
- Planning information, including budget details and associated documentation.
- Award criteria, anticipated values and tender submission information.
- Details of the consultation periods for the tender.
- Details of the number of organisations that submitted proposals and a list of their names.

[64] The OCDS is taken as a reference that defines what procurement data should be available, but there is currently a large gap between the data that is published and the data required by this standard. This does not allow the proposed method to generate OCDS-compliant procurement data sets.

- Contracting documentation with contract signature data.
- Information and milestone dates for the implementation of the contract.
- Documents related to the implementation of the contract.
- At an intermediate level it is also recommended to provide data related to awards and to identify information from each organisation making a proposal.

Continuing with the description of the method, each of the stages is made up of work with a special objective that is achieved by carrying out specific tasks. Figure 7-4 shows the work and tasks that make up the stages of the proposed method.

Figure 7-4. Jobs and tasks of the proposed method

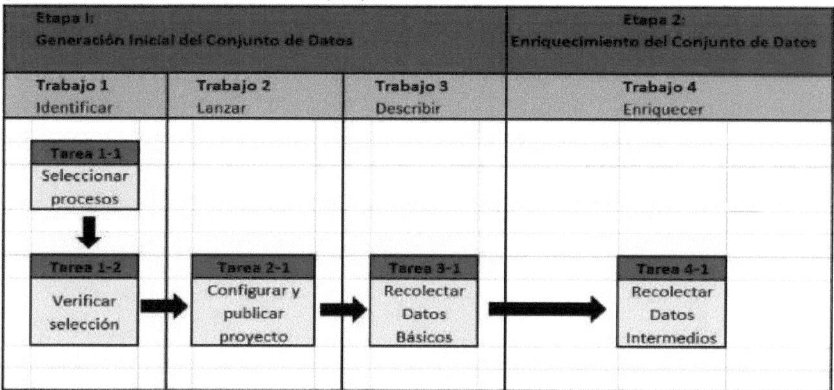

Source: Own elaboration

The first task of step 1 is called "identify" and aims to allow those interested in generating the dataset to indicate the transparency portal where the procurement data is available and to define which procurement processes are of interest.

To carry out the above, two tasks are defined: the first task (task 1-1) called "select" is precisely the one that allows stakeholders to define the transparency portal together with the criteria to be used to select the contracting processes of interest; the second task (task 1-2) called "verify" allows to confirm that the data provided in the previous task are correct and to proceed to create the crowdsourcing project to advance the data collection.

The second job in stage 1 is called "launch" and aims to allow the stakeholders or data generation project managers to configure the characteristics of the crowdsourcing project and launch or publish the project to the crowd of potential workers. Examples of characteristics to be set at this point are the number of responses (or contributions) required for the crowdsourcing tasks and the incentives to be delivered among others. The above actions are performed in task 2-1 called "set up and publish".

The last task of stage 1 (task 3) is called "describe" and is the one that allows for the collection of basic data from the crowd workers selected for this task. This is done in task 3-1 called "collect basic data".

Continuing with the description of the method, in stage 2 there is task 4 called "enrich" which seeks to allow complementing the initial dataset by adding more intermediate level data of the selected processes or processes of interest. For this purpose, task 4-1 called "collect intermediate data" is defined, where the workers in the crowd selected to carry out this task collect the indicated intermediate data.

It is important to stress that the proposed method tries to collect data defined in the OCDS at the basic and intermediate level, but in practice the transparency portals do not provide all the data required by the standard. Similarly, the method does not contemplate the collection of data defined at the advanced level of OCDS because these data are not available in the transparency portals. Data at the advanced level correspond to other data sources that are additional to the procurement process and are not necessarily publicly available on the web.

7.2.2 Detailed description of the proposed method

Following the *"Top-down"* approach, the method is presented again, first by means of an activity diagram that provides an overview of the method and the flow of steps it defines (see Figure 8-5), then in Table 8-1 the complete structure of the method is finally presented, defining elements that have been called assets, which correspond to the inputs required to carry out a certain step, and outputs, which correspond to the product that is generated as a result of carrying out a certain step.

Figure 7-5. Activity diagram of the proposed method

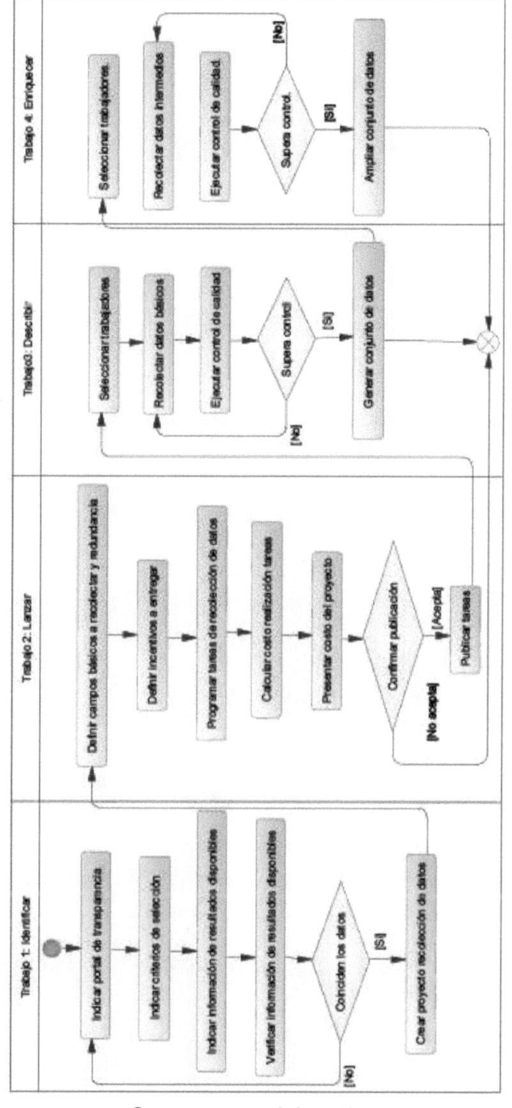

Source: own elaboration

Table 7-1. Detailed description of the proposed method

Trabajo	Tarea	Ciclo PHVA	Activos	Pasos	Salidas
Trabajo 1 Identificar	Tarea 1-1 Seleccionar procesos	P		1. Indicar portal de transparencia	(a) Url del portal de transparencia
		P		2. Indicar criterios para seleccionar procesos de contratación	(b) Lista de criterios para seleccionar los procesos de contratación de interés
		H	(a) Url del portal de transparencia	3. Ir al portal de transparencia ingresando los criterios de búsqueda definidos e indicar el número de registros (procesos de contratación) disponibles y el número de páginas que contienen los resultados	(c) Número de registros, (d) Número de páginas
	Tarea 1-2	V	(a) Url del portal de transparencia	4. Ingresar al portal de transparencia e indicar si está en funcionamiento	(e) Url del portal de transparencia verificada
		H	(b) Lista criterios de selección de procesos de contratación	5. Ingresar al portal de transparencia ingresando los criterios de búsqueda definidos e indicar el número de registros (procesos de contratación) disponibles y el número de páginas que contienen los resultados	(f) Número de registros, (g) Número de páginas

	Verificar selección	V	(c) Número de registros, (d) número de páginas	6. Verificar que el número de registros (c y f) y el número de páginas (d y g) sea el mismo	(h) Número de registros verificado, (i) Número de páginas verificado
		A		7. Crear proyecto recolección de datos	(j) Nombre del proyecto, descripción y otros datos solicitados
		P	(h) Número de registros verificado, (i) Número de páginas verificado	8. Definir los datos o campos de nivel básico e intermedio a recolectar y el número de contribuciones o respuestas requeridas	(k) Datos (campos) a recolectar y número de contribuciones (respuestas) requeridas
		P	(h) Número de registros verificado, (i) Número de páginas verificado	9. Definir el incentivo a entregar para motivar y activar la participación de los trabajadores en las tareas de recolección de datos básicos e intermedios a publicar (tareas 3-1 y 4-1)	(l) Incentivos a entregar en las tareas de recolección de datos básicos e intermedios
Trabajo 2	Tarea 2-1	H	Todos las salidas/datos (a) a (l)	10. Programar las tareas de recolección de datos básicos e intermedios (tareas 3-1 y 4-1) para el subconjunto de procesos seleccionados	(m) Planificación de las tareas de captura de datos básicos e intermedios
Lanzar	Configurar y publicar proyecto				

		H	(m) Planificación de las tareas de captura de datos básicos e intermedios	11. Calcular costo de realización de las tareas	(n) Costo del proyecto de recolección de datos
		V	(n) Costo del proyecto de recolección de datos	12. Presentar costo del proyecto de recolección de datos y solicitar confirmación	(o) Confirmación del proyecto
		A	Todos las salidas/datos (a) a (o)	13. Publicar las tareas utilizando otra plataforma de Crowdsourcing y reclutar trabajadores de la multitud	(p) Listado de trabajadores postulados para realizar las tareas del trabajo 3 y el trabajo 4

Trabajo 3	Tarea 3-1	P	(p) Listado de trabajadores postulados para realizar las tareas del trabajo 3	14. Seleccionar los trabajadores para realizar la tarea de recolección de datos básicos	(q) Listado de trabajadores seleccionados para realizar la tarea 3-1
		H	(m) Planificación de las tareas de captura de datos básicos (Urls, páginas y rango de registros asignados)	15. Acceder a las Urls y páginas indicadas e identificar los registros y datos básicos solicitados, copiar (preferiblemente siempre que sea posible) o trascribir éstos datos	(r) Datos básicos en bruto (sin agregar) de los procesos seleccionados

Describir	Recolectar datos básicos	V	(r) Datos básicos en bruto (sin agregar) de los procesos seleccionados	16. Ejecutar procedimiento de aseguramiento de calidad agregando resultados. (Las tareas que no pasen la verificación se repiten -> regresar a 15)	(s) Datos básicos de los procesos seleccionados (agregados)
		A	(s) Datos básicos de los procesos seleccionados (agregados)	17. Generar datos básicos de los procesos de contratación seleccionados incluyendo el listado de Urls donde se provee la información detallada	(t) Conjunto de datos inicial de los procesos de contratación pública seleccionados
Trabajo 4	Tarea 4-1	P	(p) Listado de trabajadores postulados para realizar las tareas del trabajo 4	18. Seleccionar los trabajadores para realizar la terea de recolección de datos intermedios	(u) Listado de trabajadores que realizaran la tarea 4-1
		H	(m) Planificación de las tareas de captura de datos básicos (Urls, páginas y rango de registros asignados)	19. Acceder a las Urls y páginas indicadas e identificar los registros y datos intermedios solicitados, copiar (preferiblemente siempre que sea posible) o trascribir éstos datos	(v) Datos intermedios de los procesos seleccionados (sin agregar)

Source: own elaboration.

Table 8-1 presents the proposed method in detail but before proceeding with the explanation of each task and its set of steps it is useful to highlight a few things:

- First, the terminology and elements used in the definition of the method are in accordance with the term crowdsourcing. The conceptual reference used to approach the method covering the core aspects of crowdsourcing was the taxonomy formulated by Hosseini et al.[65] in which four fundamental pillars are defined which are the crowd of people, the *crowdsourcer*, *the* task or activity and the platform.
- Second, the proposed method clearly defines the specifications for the four pillars of crowdsourcing according to Hosseini's taxonomy, i.e. it establishes *the* tasks (*the task*) to be performed by *the crowd* (*the crowd*) and by the interested party (*the crowdsourcer)*. Additionally, two elements are defined, the assets and outputs,

[65] HOSSEINI, M., PHALP, K., TAYLOR, J. and R. Ali. The Four Pillars of Crowdsourcing: A Reference Model, In: 8th International Conference on Research Challenges in Information Science. Marrakesh: IEEE, 2014, p.28.

associated with the tasks, which were suggested in the typology defined for crowdsourcing by Dunn and Hedges[66] as they are very convenient for the definition of the method as they allow to indicate the inputs and outputs of each task. The platform can be developed by defining the requirements to be met based on the specifications defined by the method.

• Four, the set of steps that make up the tasks are set out in accordance with the PHVA (Plan, Do, Check, Act) life cycle, understanding its application as follows: some steps of the tasks correspond to planning aspects (P), such as establishing where the information is located or what information is required, followed by doing (H), which corresponds to executing what was planned, for example, going to the source of information and collecting the required data; Finally, depending on the results of the verification, the next step is to act (A), which in this case corresponds to the generation of the desired results, such as the creation of the dataset.

To complete the presentation of the method, an explanation of the specific set of steps that make up each of the tasks is given below.

7.2.2.1 Select: Task 1-1

The purpose of this task is to allow those interested in generating the dataset to indicate the transparency portal where the data of the procurement processes are available and to define which of these processes are of interest. To do this, three steps are defined. In step (1) the interested party (*crowdsourcer*) must indicate the URL of the transparency portal where the data can be found, this value will form the output (a) and in step (2) they must indicate the criteria that will be used to select the contracting processes of interest, output (b). For example, you can be interested only in the procurement of a certain public entity, city or department, etc. The options available for selecting the processes will depend on the transparency portal.

In the last step of this task, step (3), output (a) is used as an asset and the interested party must enter the transparency portal indicated there and carry out a search using the criteria defined in output (b), which is also taken as an asset at this point (in the method, outputs generated in previous steps are used as assets). Then, based on the results obtained, the interested party must indicate some important data[67] that will be used later to programme the data collection tasks. These two data will be the outputs to be generated at this point: the number of processes (records) of interest found, output (c) and the number of web pages used to present the information of these recruitment processes including how many records are displayed per page, output (d).

In this task, steps (1) and (2) are planning (P), step (3) corresponds to doing (H). Verification (V) and acting (A) are carried out in steps of the next task.

7.2.2.2 Verify: Task 1-2

With the outputs of the previous task you could proceed to create the crowdsourcing project that will allow you to collect the data, but first you should verify that both the URL of the transparency portal (output a), the number of records (output b) and the

[66] DUNN, S., HEDGES, M., Crowd-sourcing as a component of humanities research infrastructures, In: International Journal of Humanities and Arts Computing. 2013, vol.7, no.2, p. 161

[67] Data related to the pagination of results used by websites or web applications to present information to users.

number of pages (output c) are correct.

Verification is the purpose of task 1-2 for this purpose verification step (4) (V) asks the interested party to enter the transparency portal indicated by output (a) to verify that it is operational. Then, in step (5), the interested party is asked, using the selection criteria defined in output (b) for procurement processes, to indicate again, based on the results of the search, the number of records and pages obtained as a result of the query, including how many records are displayed per page. These two values are outputs (f) and (g) which are then verified in step (6) by checking that they are equal to the values initially indicated in outputs (c) and (g) respectively. Finally, in step (7) we proceed to create the data collection project that corresponds to do (H) of the task where the output (j) is generated containing data such as the project name and description among others.

7.2.2.3 Configure and publish: Task 2-1

Once the data collection project is created, the stakeholders or project managers need to configure some features of the project before launching it or publishing it to the *crowd of* potential workers.

In order to do so, several steps are defined, starting with planning aspects (P). First, in steps (8) and (9) the stakeholder must define on the one hand the basic and intermediate level data or fields to be collected and the number of contributions or responses required (output k), and on the other hand the incentives to be given to motivate and activate the participation of workers (output l).

Subsequently, steps (10) and (11) proceed to do (H). In step (10) the basic and intermediate data capture tasks for the selected subset of processes (output m) are scheduled. In order to perform the scheduling, information generated in the previous steps such as the verified number of records (output h), the verified number of web pages presenting the results including how many records are displayed per page (output i) and the fields to be collected and responses required (output k) are used as assets.

Next, in step (11), the cost of performing the tasks (output n) is calculated using the planning information defined in the previous step (output m) as assets.

Before publishing the project in step (12) a verification (V) is performed, which consists of presenting the information of the cost of the data collection project to the interested party (*crowdsourcer*) to confirm if he/she agrees (output o), to finally proceed in step (13) to do (H), which at this point corresponds to publish the tasks using another Crowdsourcing platform seeking to recruit potential workers. The list of workers postulated to perform the tasks of job 3 and job 4 corresponds to the output (p) generated in this step.

It is important to highlight that the method defines that the publication of the data collection tasks (task 3-1 and task 4-1) are carried out on an existing crowdsourcing platform in order to facilitate the recruitment of crowd workers, due to the fact that existing platforms such as Amazon Mechanical Turk have millions of registered users. However, the performance of the published tasks for data collection is carried out using the platform that is developed for this purpose.

7.2.2.4 Collecting baseline data: Task 3-1

The first stage of the method called the initial generation of the dataset concludes with work 3 (Describe) which is intended to enable basic data collection from the

workers in the crowd who are selected to carry out this work.
To carry out the above, task 3-1 called "collect basic data" is defined, which starts with planning step (14) (P) where the workers who will carry out the collection of basic data (output q) are selected. At this point the asset to be used is the output (p) which contains the list of workers nominated to perform the task in question.
Next, in step (15), step (H) is performed in which the selected workers must access the URLs and pages indicated to identify the recruitment processes (records) and basic data requested, which they must copy (preferably whenever possible) or transcribe if the conditions of presentation of the information do not allow it (e.g. in scanned documents). This step uses the asset (m) which is one of the most important in the whole process because it contains the planning and layout generated to carry out the basic data capture. Also, the output (r) generated in this part corresponds precisely to the raw (unaggregated) basic data collected from the processes of interest.
Subsequently, in step (16), verification (V) is performed, which corresponds to executing the defined quality control procedure based on the aggregation of results. The asset to be used is the output (r) containing the basic raw data which are aggregated to generate the output (s) containing the basic data but this time aggregated, where only one value is available for each of the requested fields or variables; the basic data that do not meet the established quality requirements must be collected again by repeating the task.
The task ends with step (17) where, using the aggregated data (output s), we proceed to step (A), which corresponds to the initial generation of the basic data set of the selected procurement processes (output t), where we have the list of URLs where the detailed information of the processes of interest can be consulted.

7.2.2.5 Collect Intermediate Data: Task 4-1

Task 4-1 "collect intermediate data" is located in the second stage of the method called "dataset enrichment" where you have a single job called "enrich" that seeks to allow you to complement the initial dataset by adding more intermediate level data from the selected processes or processes of interest.
To do this, task 4-1 is defined, which sets out a similar process to the one carried out in basic data collection, but this time focusing on intermediate level data collection.
The process starts with the planning step (18) (P) where the workers who will carry out the intermediate level data collection (output u) are selected. At this point the asset to be used is the output (p) which contains the list of workers nominated to perform the task in question.
Next, in step (19), the next step is to do (H) in which selected workers must access the URLs indicated to identify the requested data, which they must copy or transcribe. In this step the asset (m) is used again, which as mentioned above contains the planning and distribution generated to carry out the capture of the intermediate level data. The output (v) generated in this part corresponds to the raw (unaggregated) intermediate level data collected from the processes of interest.
Subsequently, in step (20), verification (v) is performed, which corresponds to executing the defined quality control procedure based on the aggregation of results. The asset to be used is the output (v) which contains the raw intermediate level data that must be aggregated to generate the output (w) which contains the aggregate of

the data where only one value is available for each of the requested fields; data that do not meet the established quality requirements must be captured again by repeating this task.
The task ends with step (21) where using the aggregated data (output w) we proceed to step (A) which corresponds to perform the enrichment of the initial dataset by adding the collected intermediate level data.

CHAPTER 8

8. CROWDSOURCING APPLICATION PROTOTYPE DEVELOPED

This section describes the crowdsourcing prototype application that was created in order to test some elements of the proposed recruitment data collection method. The software development methodology taken as a reference was Iconix[68] , making a fairly complete design of an application to support the entire proposed data collection method.

For the implementation of the prototype, only the requirements defined for user registration, Crowdsourcing data collection project creation, data collection and quality control were selected in order to be able to perform some important tests in this research.

The complete artefacts generated in the development process describing the application can be found in Annex A. In this section of the document, selected artefacts are presented in order to give an idea of the system.

8.1 Analysis and requirements stage

8.1.1 Definition of System Actors

8.1.1. 1Crowdsourcer

This user of the system is interested in generating structured datasets containing information on the procurement processes of interest. The system allows you to create crowdsourcing projects through which the data collection process will be carried out and once the data collection is completed, the data can be downloaded or published so that they are available to interested citizens.

8.1.1.2 Crowdworker

This system user is the one who can perform the crowdsourcing tasks of data collection. It is a person who is part of the multitude of Web users interested in participating and working in this type of initiatives motivated by some kind of reward or benefit.

8.1.1.3 Citizen

This system user can access available public procurement datasets.

8.1.1.4 Administrator

This is the super user empowered to perform all system administration tasks related to crowdsourcing projects and users.

8.1.2 Description of functional requirements

Table 8-1 shows as an example some of the defined requirements that were selected to guide the implementation of the prototype; the original code assigned to each requirement in the design process is retained.

Table 8-1. Some requirements for the prototype application

No / VISIBILITY	DESCRIPTION
	User management
R11 / Visible	When a user wishes to register in the system, the system will ask for the following information:

[68] ICONIX is a methodology for software development that defines a set of object-oriented methods covering the entire life cycle of a project.

	• Full name • Sex • Date of birth • Telephone contact (landline or mobile) • Country • User type • Mail (e-mail) • Confirmation Mail (e-mail) • Password • Password confirmation All fields are mandatory, the date must be in day/month/year format, there are three types of user (crowdsourcer, crowdworker and citizen), the email must be in the form defined for an email (xxx@yyy.zz) and the password must have a minimum length of 8 characters combining alphanumeric characters. In addition, the email and password fields must match their respective confirmation fields.
	Crowdsourcing Project Management
R18 / Visible	When a crowdsourcer user is making the selection of the procurement processes of interest, the system will request the following information related to the source, transparency portal, where the public procurement information is located: • Url of the transparency portal where information on public procurement processes is available. • Checklist to indicate the criteria to be used to select the procurement processes of interest In Colombia, the transparency portal, a website that is part of SECOP (Electronic Public Procurement System), has the following criteria available to select the contracting processes: Type of contracting, Department of execution, Date From, Date To, Amount, Process Number, State, Municipality, Results per Page (in order to use this application in other countries, this entry must be adjusted according to the criteria existing in the country). The system will allow you to select among the existing criteria in Colombia and you must enter at least two of these criteria.
R23 / Visible	When a crowdsourcer wants to create a new crowdsourcing data collection project in the system, the system will ask for the following information: • Name of the project • Description • Deadline • List of keywords All fields are mandatory. The description has a maximum length of 600 characters, the deadline will be in the format day/month/year, up to five keywords not exceeding 50 characters together are allowed. The creation of the project is only done if the data collection

	crowdsourcing task scheduling data verification was successful.
R27 / Visible	When a crowdsourcer user wishes to define the required redundancy of responses for the basic level fields to be collected in the project, the system will prompt the user: - Number of contributions per field The field is mandatory and the value to be entered is of integer type.
R29 / Visible	When a crowdsourcer user wants to define the minimum accuracy required for the data to be collected in the project, the system will prompt him/her: - Minimum accuracy The field is mandatory and the value to be entered is selected from a
	list with the following options: 70, 80, 90.
R30 / Visible	When a crowdsourcer user wants to define a financial incentive to be given to motivate user participation, the system will prompt him/her: • Value to be paid for the collection of basic level fields per contracting process • Value to be paid for the collection of intermediate level fields per procurement process The field is mandatory if no other type of incentive is defined and the value to be entered is double. The currency used is Colombian pesos.
	Data collection
R44 / Visible	When a crowdworker user wishes to enter the basic level data collected from a recruitment process, the system will prompt the user: • Buyer information fields: entity name, address, city, town, postcode, region, country, telephone, website, email, contact name. • Fields with information on the call for tenders and contracting process: type of process, status of the process, contracting regime, group (code and description), segment (code and description), family (code and description), class (code and description), detail and quantity of the object to be contracted, amount to be contracted, type of contract. • Fields with information on the award of the process: name or company name of the selected bidder, definitive qualification, date and time of the award hearing. • Fields with information about the process schedule: date and time of opening of the process, date and time of closing of the process. Of the above fields, only those that have been defined by the user as fields of interest in the project configuration shall be available for data entry.
R46 / Visible	When the system runs the quality control mechanism it must calculate the final data for the recruitment process resulting from the aggregation and consensus of the raw data collected. The quality control mechanism corresponds to an available evolutionary algorithm which calculates the average string of the

	supplied inputs.

Source: own elaboration

8.1.3 Description of non-functional requirements

Table 8-2. Some non-functional requirements of the prototype application

No	Description
RNF1	The application must be developed with Java Server Faces (JSF 2.2) and PosgreSQL 9.5.
RNF2	The application to be developed must correspond to a web application (WebApp) under Client/Server architecture.
RNF3	The http protocol must be used for communication between the clients and the server in order to avoid problems with firewalls on the Internet.

Source: own elaboration

8.1.4 System use cases

The use case diagrams of the system are shown below, grouped according to their affinity. Subsequently, a description of some use cases with their corresponding user interface model is given.

Figure 8-1. Use case diagram for user management

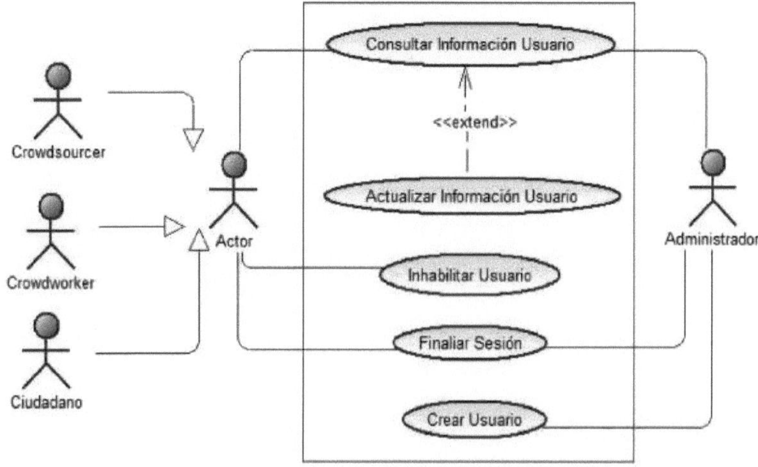

Source: Own elaboration

Figure 8-2. Use case diagram for crowdsourcing project management

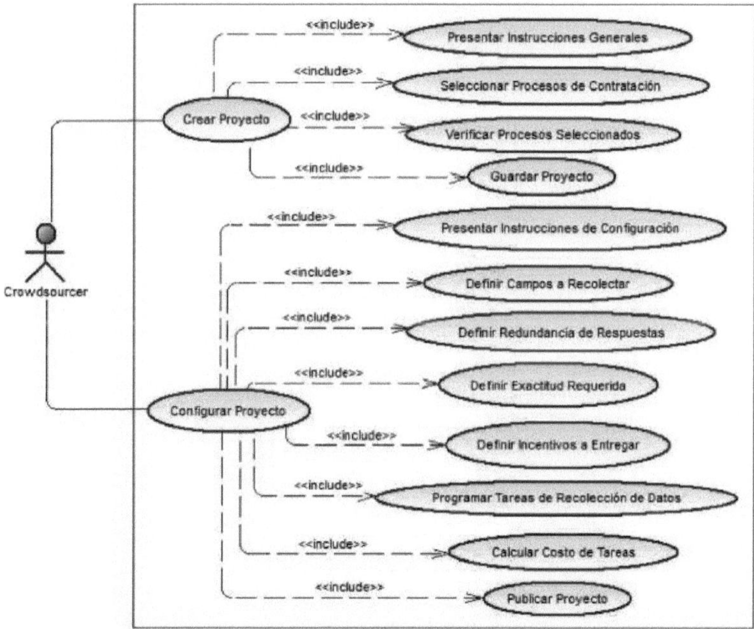

Source: Own elaboration

Figure 8-3. Use case diagram for data collection

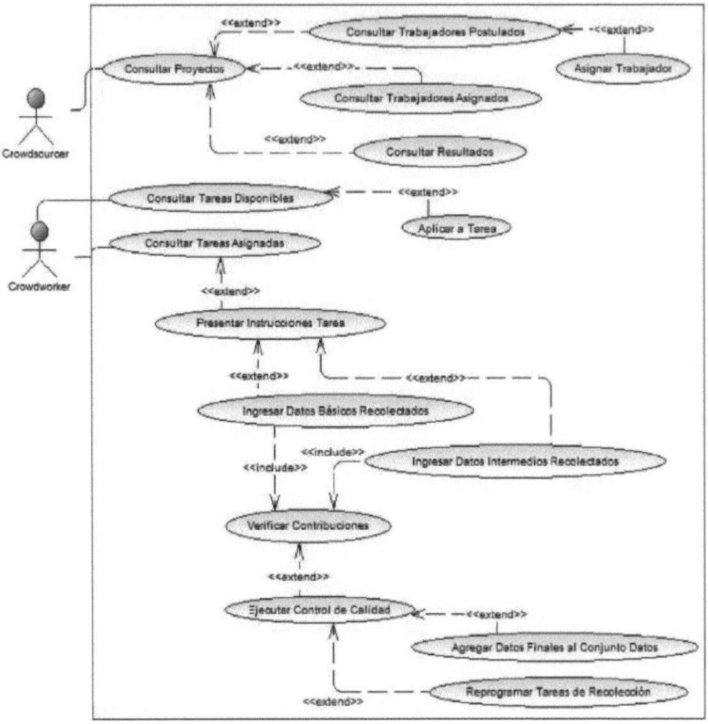

Source: Own elaboration

8.1.1. 1System use cases and graphical interface models

Table 8-3. Use case register user

GENERAL INFORMATION	
Name:	Register User.
Actors:	Crowdsourcer, crowdworker, citizen
Purpose:	Allow users to register in the system.
Preconditions:	
Post-Conditions:	User information is stored in the system.
Cross-referencing:	Functional Requirement: R11, R16.
NORMAL FLOW	
User Action	**System Response**
	1 Display the registration user interface (see figure 9-4) where the following data must be entered: • Full name • Genre • Date of birth • Telephone contact (landline or mobile) • Select country • Select user type • Electronic mail (e-mail) • Confirmation Mail (e-mail) • Password • Password confirmation

2 Enter the requested data	3 Validate that the values entered comply with the following restrictions: all fields are mandatory, the date must comply with the format day/month/year, the e-mail must have the form defined for an e-mail (xxx@yyy.zz) and the password must have a minimum length of 8 characters combining alpha-numeric characters. In addition, the email and password fields must match their respective confirmation fields. Users cannot register as administrator type.
	4 Store the data provided.
	5 Send an e-mail to the user notifying the registration in the system and indicating how to enter the system (user name to be used corresponds to the e-mail address and password to the password provided).
	6 Display a response message indicating that the data was stored.
EXCEPTIONAL FLOW	
4 If problems are encountered when validating the data provided, the corresponding response message is displayed and you return to 2. 5 If problems occur when storing the supplied data, the corresponding response message is displayed and you return to 2.	

Source: Own elaboration

Figure 8-4. Graphical user interface for registration

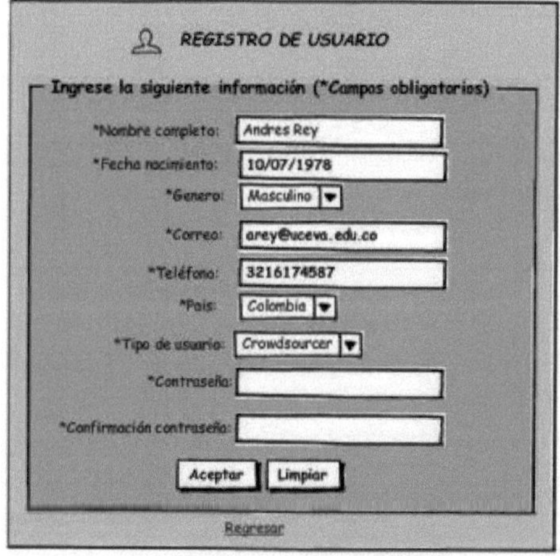

Source: Own elaboration

Table 8-4. Use case for selecting recruitment processes.

GENERAL INFORMATION	
Name:	Select Recruitment Processes.
Actors:	Crowdsourcer.
Purpose:	Allow the crowdsourcer user to enter the information required to select the procurement processes of interest for the data collection project.
Preconditions:	The user has a valid session in the system.
Post-Conditions:	The url of the transparency portal, the criteria for selecting the processes of interest, the number of processes found, the number of pages displaying this information and the number of records per page are stored in the system.
Cross-referencing:	Functional Requirement: R18, R19.
NORMAL FLOW	

User Action	System Response
	1 Display the user interface that allows you to select the procurement processes (see figure 9-6), in this interface you can enter: - Url of the transparency portal where the information on the processes of
	recruitment • Checklist to indicate the criteria for selecting the processes of interest. The available options are contracting modality, department of execution, date from, date to, amount, process number, status, municipality, results per page. • Number of processes found according to the criteria • Number of pages displaying (containing) information on the processes encountered • Number of records per page
2 Enter the requested data.	3 Validate that the values entered comply with the following restrictions: all fields are mandatory and at least two selection criteria must be indicated; the number of processes found, number of pages containing the results and the number of records per page must be positive integer values.
	4 Store the data provided.
	5 Display a response message indicating that the information has been stored.
	6 Continues with the use case from where it was invoked.
EXCEPTIONAL FLOW	
3 If problems are encountered when validating the data supplied, the corresponding response message is displayed and the system returns to 1. 4 If problems occur when storing the data supplied, the corresponding response message is displayed and the system returns to 1.	

Source: Own elaboration

Figure 8-5. Graphical interface for recruitment process selection

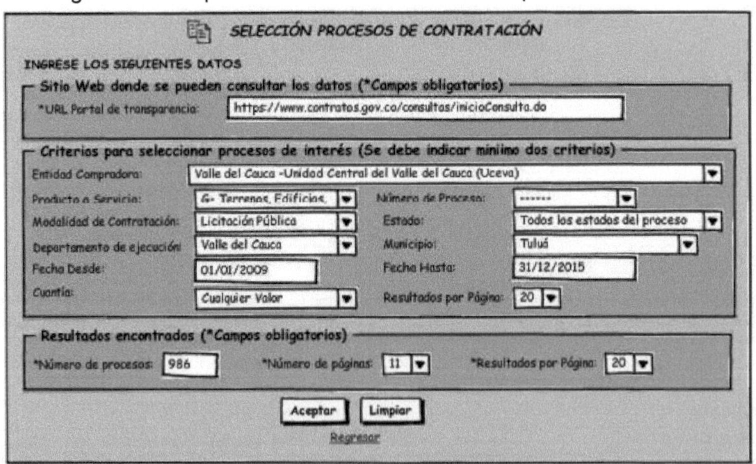

Source: Own elaboration

Table 8-5. Use case save project

GENERAL INFORMATION	
Name:	Save Project.

Actors:	Crowdsourcer.
Purpose:	Allow the crowdsourcer to save the data collection project by providing name and description.
Preconditions:	The user has a valid session in the system, the verification of the recruitment process selection data was successfully performed.
Post-Conditions:	The crowdsourcing data collection project is created.
Cross-referencing:	Functional Requirement: R23
NORMAL FLOW	
User Action	**System Response**
	1 Display the user interface for project saving (see figure 9-7) where you can login: • Name of the project • Description • Expected completion date • Keywords associated with the project
2 Enter requested data	3 Validate that the values entered comply with the following restrictions: all fields are mandatory, the description
	has a maximum length of 600 characters, the deadline will be in the format day/month/year, up to five keywords not exceeding 50 characters together are allowed.
	4 Store the supplied data of the new project.
	5 Display response message indicating that the project was saved.
	6 Continues with the use case from where it was invoked.
EXCEPTIONAL FLOW	
3 If problems are encountered when validating the data supplied, the corresponding response message is displayed and the system returns to 1. 4 If problems occur when storing the data, the corresponding response message is displayed and continue in 6.	

Source: Own elaboration

Figure 8-6. Graphical interface to save the project

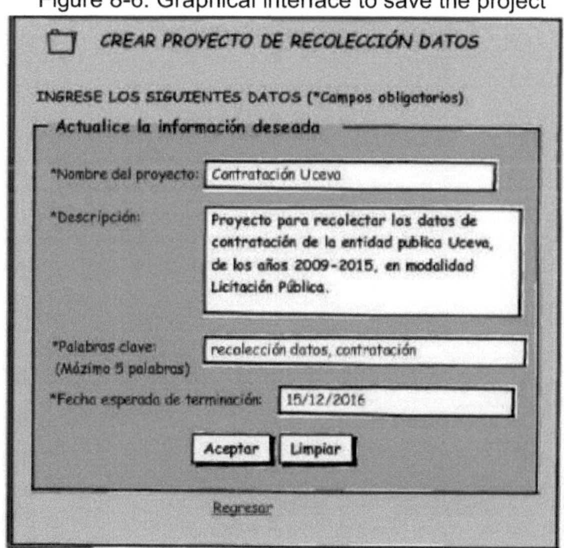

Source: Own elaboration

Table 8-6. Use case enter basic data collected.

GENERAL INFORMATION	
Name:	Enter Basic Data Collected.
Actors:	Crowdworker
Purpose:	Allow the crowdworker user to enter data collected from a recruitment process.
Preconditions:	The user has a valid session in the system, there is a data collection task to be performed for a given project.
Post-Conditions:	Basic data collected from a recruitment process are stored in the system.
Cross-referencing:	Functional Requirement: R44.
	NORMAL FLOW
User Action	**System Response**
	1 Display a user interface (see figure 9-10) that allows the following data to be entered: • Buyer information fields: entity name, address, location, region, country, telephone, website, email, contact name. • Fields with information on the call for tenders and contracting process: type of process, status of the process, contracting regime, group (code and description), segment (code and description), family (code and description), class (code and description), detail and quantity of the object to be contracted, amount to be contracted, type of contract. • Fields with information about the award of the process: nit and name or company name of the selected bidder, final score (points), country of location of the contractor, region location of the contractor, locality location of the contractor, physical address location of the contractor, name of the legal representative of the contractor, identification of the legal representative of the contractor. • Fields with information on contracts associated with the process: contract number, contract status, type of termination, object of the contract, final amount of the contract, percentage of advance payment, identification of the contractor, value of the external audit contract, date of signature of the contract, date of start of execution of the contract, execution period of the contract, destination of the expenditure, origin of the resources. • Fields with information on the timetable of the process: date and time of the opening of the process, date and time of the closing of the process, date and time of the award hearing. **Note: Data availability depends on the country and the GUI can be adjusted, e.g. in the location of input controls, to facilitate the data collection process.**
2 Enter the requested data	3 Store the data collected for the given recruitment process.
	4 Include the use case to verify contributions.
	5 Continues with the use case from where it was invoked.
EXCEPTIONAL FLOW	
3 If there are problems when storing, the corresponding response message is displayed and you continue in 5. message is displayed and continue in 5.	

Source: Own elaboration

Figure 8-7. Graphical interface to basic level data collection

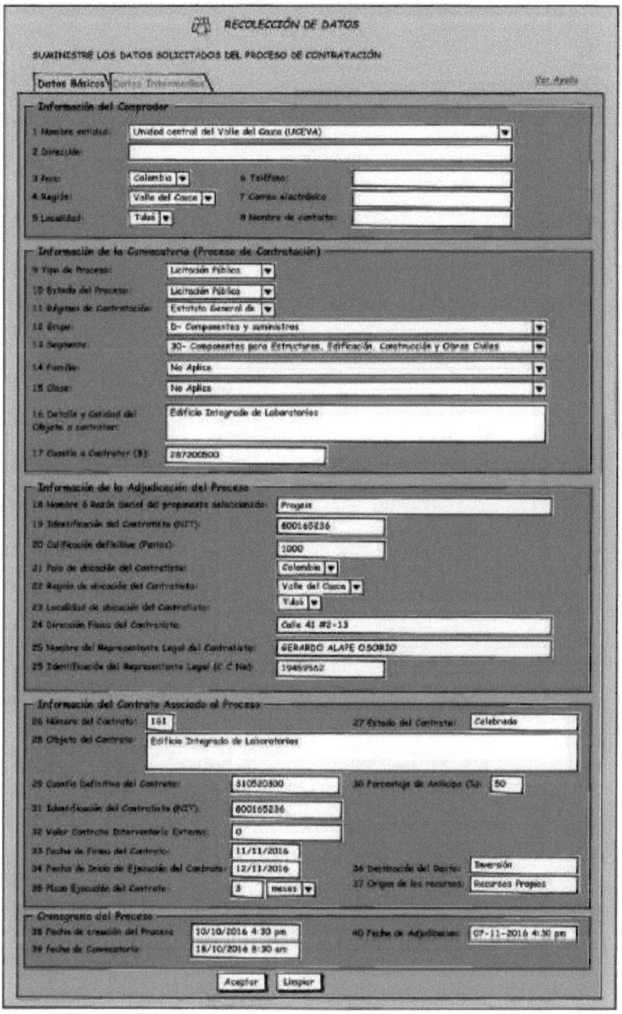

Source: Own elaboration

8.2 Design stage

8.2.1 Class diagrams

Below are several diagrams showing the classes and interfaces defined.

Figure 8-8. Class diagram part 1

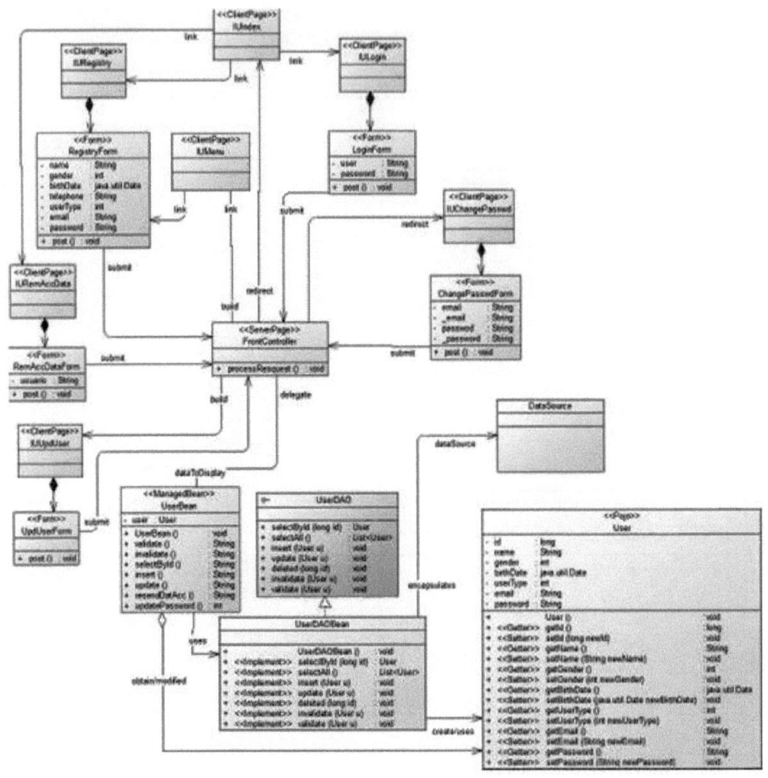

<<ClientPage>>
IUIndex
IURegistry
IULogin
<<Form>>
RegistryForm
IUMenu
LoginForm
IUChangePassword
IURemAccData
ChangePasswordForm
<<ServerPage>>
FrontController
processRequest ()
RemAccDataForm
IUUpdUser
UpdUserForm
DataSource
<<ManagedBean>>
UserBean
UserDAO
UserDAOBean
<<Pojo>>
User
link
submit
redirect
build
delegate
dataToDisplay
dataSource
encapsulates
uses
obtain/modified
create/uses

Source: Own elaboration

Figure 8-9. Class diagram part 2

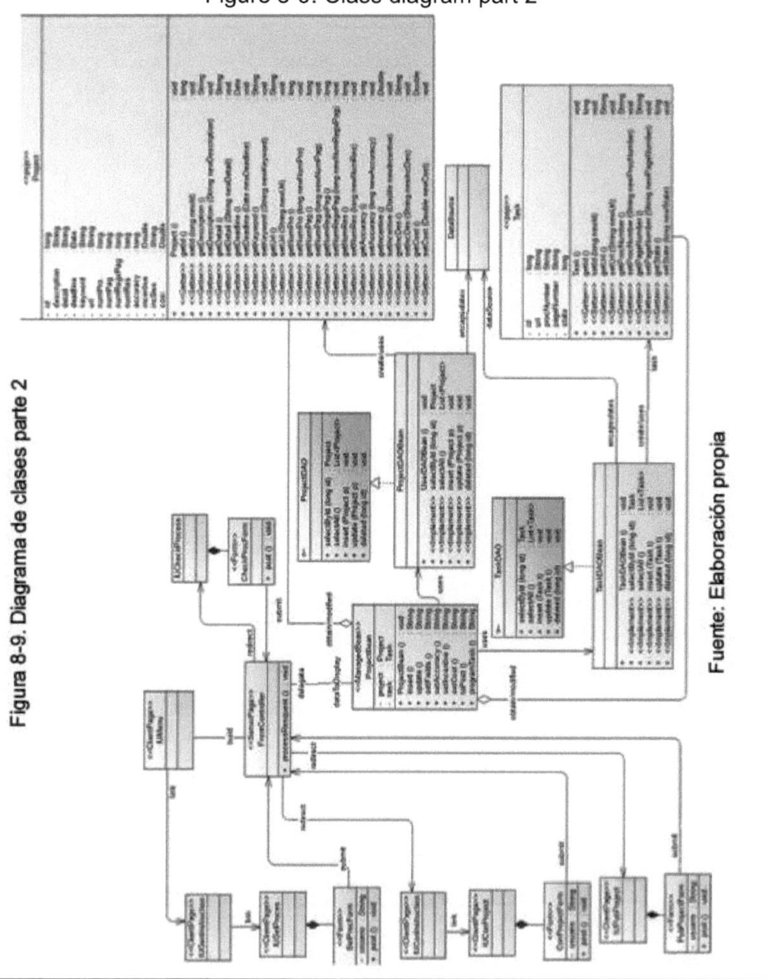

Source: Own elaboration

8.2.2 Database model

Figure 8-10. Database model

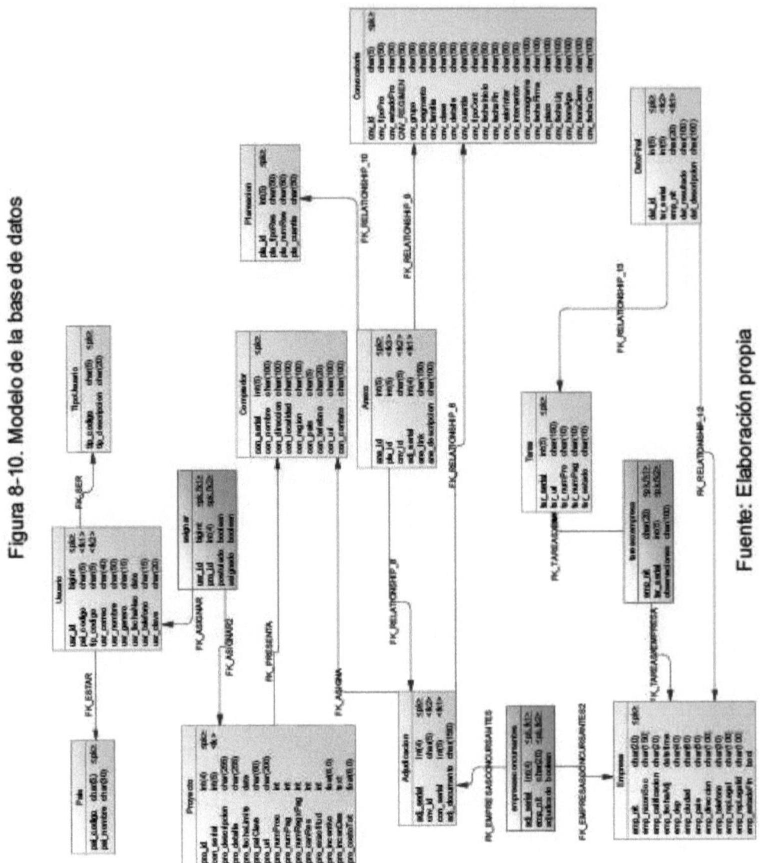

Source: Own elaboration

8.3 Implementation stage

8.3.1 Programming languages and tools used

Table 9-9 describes the different technologies and tools used in the development of the prototype application.

Table 8-7. Technologies and tools used in development.

Purpose	Technology / Tool
Programming language	Java
Web development framework	Java Server Faces (JSF 2.2) / Primefaces
IDE	NetBeans 8.2
Application Server	Glassfish 4.1
Database server	PosgreSQL 9.5

Source: own elaboration.

8.3.2 Deployment diagram

Figure 8-11. Deployment Diagram

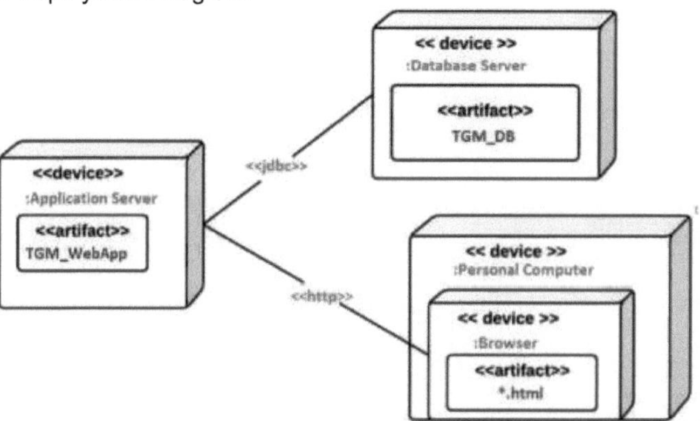

Source: own elaboration

8.3.3 Prototype Application Implemented

The source code of the different classes that make up the prototype is available in Annex B. The implemented prototype allows to register users, create Crowdsourcing data collection projects, advance the collection of basic data and perform quality control with the approach of aggregation of results through the evolutionary algorithm described in section 10. To conclude this section, some graphical interfaces of the developed prototype application are shown.

Figure 8-12. Start-up guide

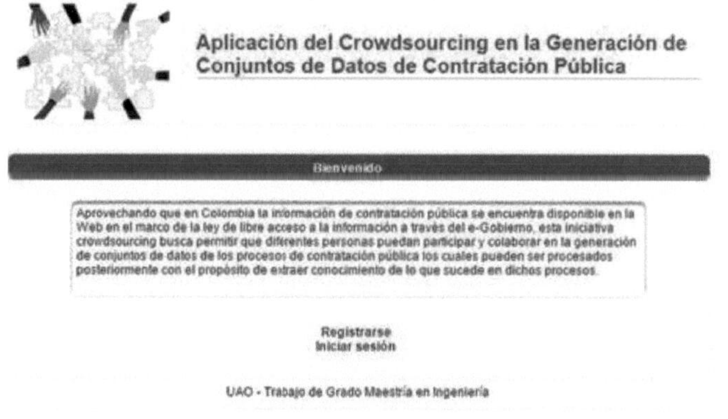

Source: own elaboration.

Figure 8-13. Login Guide

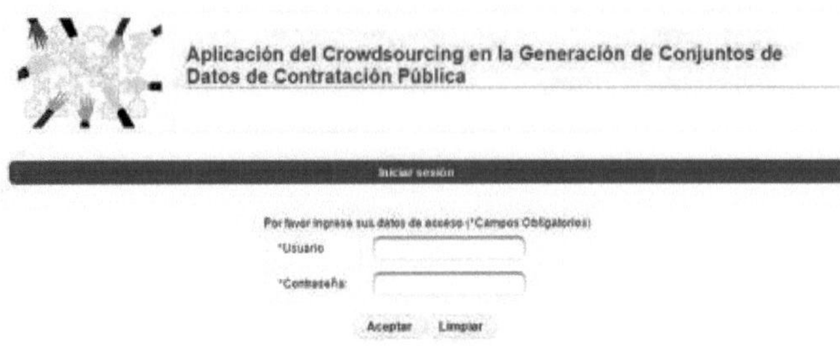

Source: own elaboration.

Figure 8-14.User Registration Guide

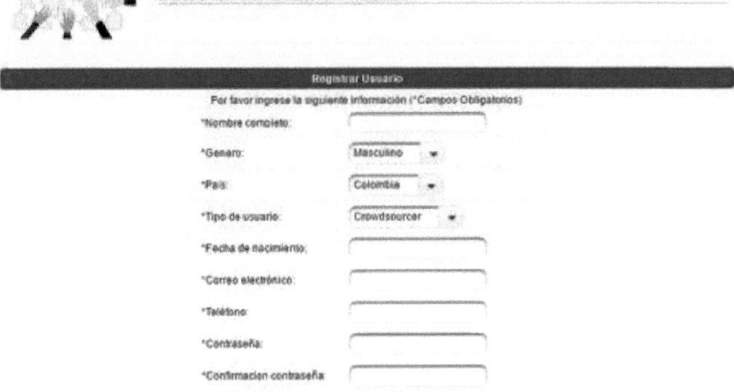

Source: own elaboration.

Figure 8-15. Project Creation Guide

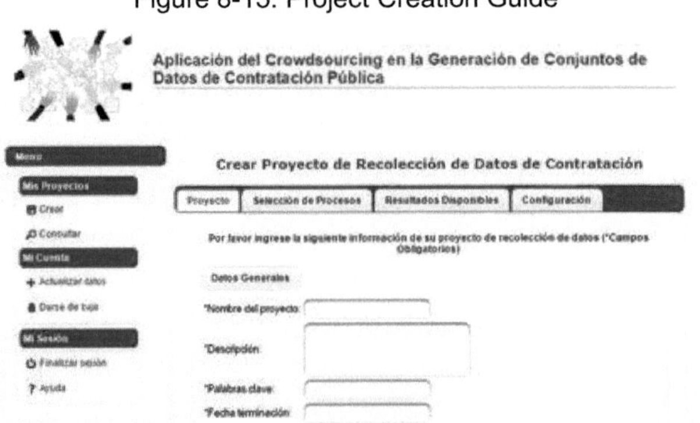

Source: own elaboration.

Figure 8-16. Process selection guide

Proyecto | Selección de Procesos | Resultados Disponibles | Configuración

Por favor ingrese la siguiente información requerida seleccionar los procesos de contratación de interés (*Campos Obligatorios)

Portal de Trasparencia

*URL del Portal de transparencia:

Criterios para seleccionar los procesos de contratación de interés

*Entidad compradora: Unidad Central del Valle del Cauca
*Producto o servicio:
*Número de proceso:
*Modalidad de contratación: Licitación Pública
*Estado del proceso: Todos
*Departamento de ejecución: Valle del cauca
*Ciudad de ejecución: Tuluá
*Fecha desde:
*Fecha hasta:
*Cuantía ($):

Source: own elaboration.

Figure 8-17. Available results guide

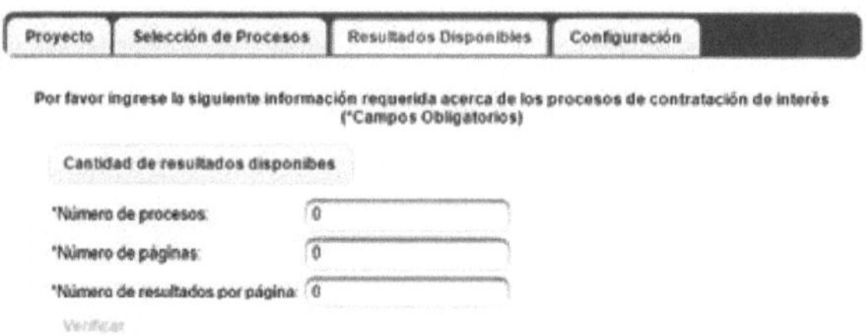

Source: own elaboration.

Figure 8-18. Project configuration guide

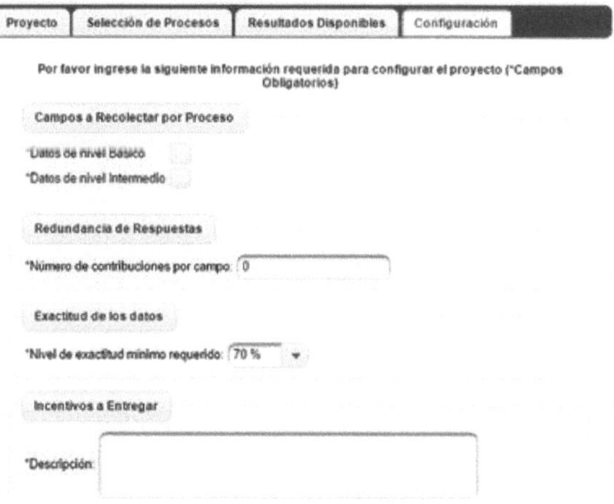

Source: own elaboration.

CHAPTER 9

9. MECHANISM DEVELOPED TO GENERATE FINAL DATA

Data collected using the crowdsourcing concept is potentially noisy, i.e. it may contain errors that affect its accuracy, in this work such data is referred to as "raw data".

This section describes a developed mechanism proposed for the correction of errors introduced in data collected through Crowdsourcing. The mechanism in question allows obtaining final data resulting from the aggregation and consensus of the raw data.

The session is structured as follows: first, it describes the approach adopted for error correction based on the aggregation of results, which is one of the most promising mechanisms used in crowdsourcing to deal with the quality of the results; second, it mentions a complex problem called *String Media* calculation encountered when trying to perform the aggregation of the character strings of the raw data; third, it describes the implemented solution to the mentioned problem which is based on evolutionary programming.

9.1 Aggregation of results in crowdsourcing

The data obtained through the application of the crowdsourcing concept may contain errors, to address this problem we use the aggregation of results that generates redundancy with several answers for the same task provided by different workers, this allows us to obtain a more reliable result.

When approaching the development of a solution that would apply the concept of aggregation of results to generate final data, it was found that aggregating the character strings that constitute the raw data is not trivial at all, and in the literature this problem is referred to as the average string calculation described below.

9.2 Problem of calculating the average chain

In section 6.2.7 the definition of the chain was presented, but although it is apparently simple, de la Higuera[69] showed that the exact calculation of the chain is a NP-hard problem *(Non-Deterministic Polynomial-Time Hard)*, so in practice approaches and algorithms must be used that find the approximate mean chain with a polynomial cost and not an exponential cost as when the exact mean chain is found.

At this point in the work, the problem to be solved focused on developing a solution to find a good approximation to the average string of a set of given character strings. The solution developed is detailed below.

9.3 Solution to the problem of finding the average chain by means of evolutionary algorithms.

The solution to the problem of finding the average chain was carried out by means of evolutionary algorithms using the genetic programming approach. This type of programming[70] was selected taking into account that evolutionary algorithms are used, with very good results, in problems that are computationally difficult to solve, i.e. those that do not have a solution in polynomial time. The evolutionary algorithm

[69] DE LA HIGUERA, Colin; CASACUBERTA, Francisco. Topology of strings: Median string is NP-complete. In: *Theoretical computer science*, 2000, vol.230, no 1-2, p.39.

[70] There are also other types of options such as voracious programming, dynamic programming, AI (artificial intelligence) searches, neural networks, among others that can be explored.

developed and the results of the tests carried out are described below.

9.3.1 Evolutionary algorithm calculating the average chain

To begin the description of the evolutionary algorithm, Figure 10-1 is provided, which corresponds to the class diagram of the implemented solution.

Figure 9-1. Class diagram evolutionary algorithm

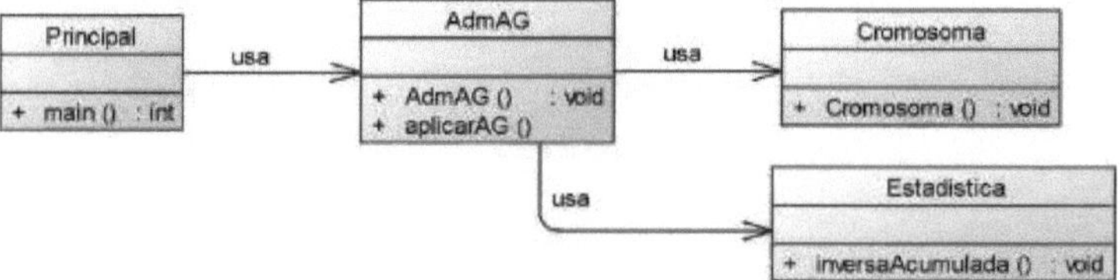

Source: own elaboration

As shown in the figure above, there are four classes:

- Main Class in charge of initiating the execution of the evolutionary algorithm.
- A class called AdmAG that implements a series of steps defined in evolutionary programming such as initialisation, evaluation, selection, replication and replacement.
- Chromosome class used to model the chromosomes that are the basic elements in the modelling proposed by evolutionary algorithms.
- Statistical class containing methods to support various statistical operations commonly used in evolutionary algorithms using the sampling technique for the selection of the best chromosomes.

The four previous classes allow to find the average string and the process starts when the main class creates an instance of the evolutionary algorithm class (AdmAG) passing as parameters the set of strings to which the average string is going to be calculated. The input strings correspond to raw data that are loaded by the main class from an input file.

When creating an instance of the evolutionary algorithm class, the constructor method is responsible for creating the table of symbols to be used, including only the characters that appear in the input strings passed as arguments, in order to optimise the search space. Once this is done, the applyAG method is invoked, which is in charge of executing the steps shown in Figure 10-2 defined in the evolutionary computation.

Figure 9-2. Steps followed in evolutionary computation

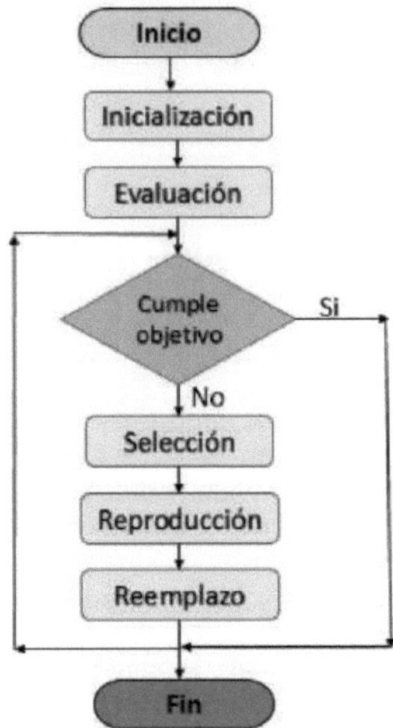

Source: own elaboration

To understand how the evolutionary process of creating and exploring a large search space[71] takes place with the aim of finding the average chain, a description of what happens at each step is presented:

- (1) Initialisation: It consists of filling the warehouse with 100 chromosomes. The warehouse is implemented as a vector of chromosomes basically made up of random strings created according to the symbol table. Each chromosome represents a possible solution to the mean chain problem.

(2) Evaluation: This consists of ordering the chromosomes in the cellar from smallest to largest. Figure 10-3 attempts to represent the cellar and how the chromosomes are organised according to the evaluation value, which is obtained by comparing the current string (which makes up the chromosome) with respect to the input strings received. The evaluation value represents how close a chromosome is to the input strings. The evaluation values are obtained for each chromosome at the time of sorting and are stored as part of its structure.

Figure 9-3. Representation of the chromosome cellar

[71] The search space is made up of chromosomes containing strings created from the input strings. It is such a large space of string combinations that the task is akin to "looking for a needle in a haystack".

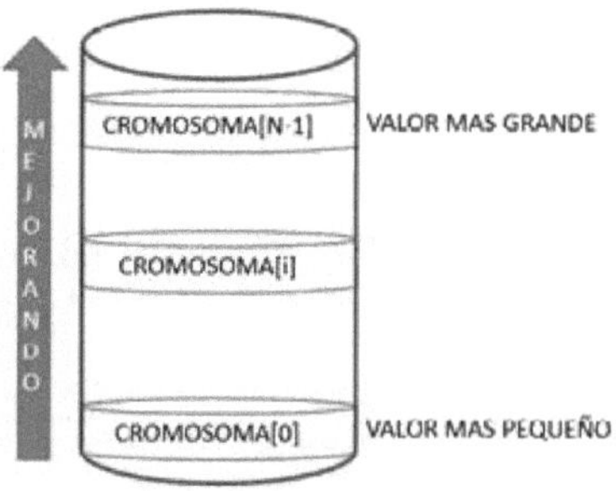

Source: own elaboration

(3) Selection: With the evaluation values, a vector is created that is passed as a parameter to create an instance of the Statistics class, which is in charge of choosing the statistically most suitable chromosomes according to their evaluation value. It is important to emphasise that before creating this instance, negative numbers or zeros must be avoided, so before using the statistics class, the minimum of the evaluation function is calculated and if this value is not positive, a shift or displacement is made, that is, a kind of pre-normalisation is made because the statistics class makes the real normalisation in order to apply the accumulated inverse function (main function of the class) which is in charge of randomly choosing the fittest chromosomes that are copied to a vector called "matingPool". Figure 10-4 tries to represent how the chromosomes with the highest evaluation value are more likely to be chosen, but all of them have some probability including the least fit ones, because for their selection in the cumulative inverse function a strategy called sampling draw[72] is used; the amount[73] of chromosomes that are passed to the matingPool corresponds to 10% of the size of the pool.

Figure 9-4. Representation of the mating pool

[72] There are also other strategies that can be used to select chromosomes such as: selecting the first n most suitable chromosomes, by lottery, by roulette, by leftovers, by tournament, etc. The selection of any of these strategies impacts how much of the search space is explored.

[73] The recommended value for the matingPool size is generally 10% but can be adjusted depending on the problem and results of tests performed on the algorithm.

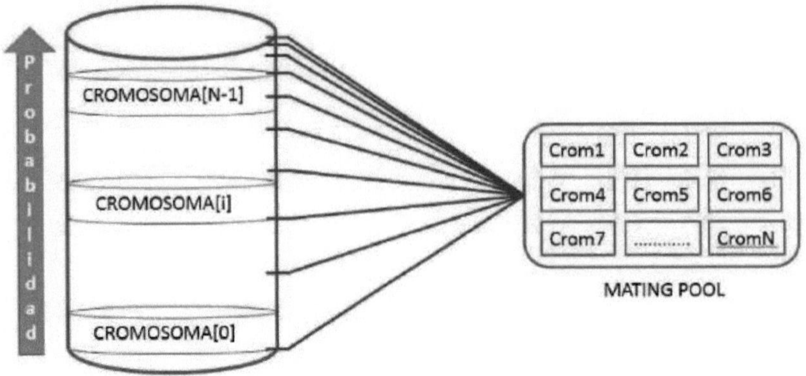

Source: own elaboration

- (4) Reproduction: Already in the Mating Pool, two chromosomes are chosen that are located next to each other and a decision is made at[74] as to whether both will mutate or crossover with each other. If they mutate, it is decided randomly which of the four defined mutation functions is to be applied and if they are to crossover, it is decided randomly which of the two crossover functions is to be chosen. The new chromosomes resulting from the mutation and/or crossover are taken to a new chromosome vector and when all the reproductions are finished and the vector is complete, it is sorted according to the evaluation function; the reproduction ends when all the chromosomes are taken from the Mating Pool.
- (7) Replacement: To replace chromosomes in the cellar with the new chromosomes resulting from the reproduction, the criterion is used: if it is better than the worst chromosome in the cellar, then it enters the cellar. This implies that for one of the new chromosomes to enter the cellar, its evaluation function must be better than that of the worst chromosome in the cellar; new chromosomes that do not enter the cellar are discarded because they are not "better" than the existing chromosomes. After introducing the new chromosomes into the cellar, the cellar is sorted and the process is started again until it is observed that no new chromosomes are entering the cellar, so the algorithm is stopped and the highest chromosome in the cellar is returned, which will be the one that contains the average chain that was desired to be found.

9.3.2 Representation of individuals

In the evolutionary model defined for the medium chain problem, the chromosome is made up of a single gene that corresponds to a string of characters (see Figure 10-5). Each chromosome represents a possible solution to the medium string problem and when creating the strings that make it up, the following considerations are taken

[74] Generally, the percentage of mutation is very low compared to that of crossover, i.e. 95% corresponds to crossover and 5% to mutation. For the mid-chain calculation problem the percentages defined were 50% for both crossover and mutation because the operators defined in the Levenshtein distance are mutational, which is very different from other problems where a high mutation could affect and damage the chromosomes.

into account: the sizes of these strings range from the minimum size of the input strings to the largest size of the input strings; all the characters are chosen from the symbol table that is created only using characters that appear in the input strings; the above restrictions correspond to heuristics that are totally logical since the medium string cannot be smaller than the minimum, nor larger than the maximum.

Other fields used to store data are also defined in the chromosome, but it is the evaluation value and the possible average string that are the most important.

Figure 9-5. Representation of the chromosome in the mid-chain problem.

Source: own elaboration

9.3.3 Generation of the initial population

Traditionally, it is recommended to start testing evolutionary algorithms using 100 chromosomes as the starting population in the cellar; this value can be varied depending on the problem and the results obtained in the experimentation. In the algorithm developed, this guideline was followed and it was not necessary to test with other values.

9.3.4 Evaluation Function

The evaluation function allows finding a value that represents how close the chain that conforms the chromosome is to the average of the input chains. That is to say, how similar or similar is this chain with respect to the set of input chains that remain unaltered.

The evaluation function applied to measure the fitness of chromosomes in the evolutionary algorithm is as follows (equation 2):

Assessment Value = Similarity by Column - Total Dissimilarity (2)

The terms used in the evaluation are described below:

- Similarity per column. It corresponds to the evaluation of the similarity of the characters of the evaluated strings in each column or position in the following way: if the characters of the column are the same, a value of 1 is assigned, if they are different -1 and if it is a blank space -2. The final value of the evaluation of the similarity per column corresponds to the sum of the results of the evaluation of each column when the string that conforms the chromosome is compared against all the input strings. A high evaluation value represents that the strings are similar, but it may be the case that negative values are obtained, representing that the strings are not similar.
- Total dissimilarity. This measure represents how different the evaluated strings are. Several methods can be used to measure the dissimilarity of strings but the most widely used is the Levenshtein method that allows finding a value called edit distance, which represents how different the strings are. When the edit distance is zero it means that the strings are equal, but as it increases it represents greater

dissimilarity. The total dissimilarity measure is always positive.

9.3.5 Crossover and mutation operators

The functions or operators defined to apply to the chromosome at the time of reproduction are as follows.

- Crossover. There are two ways:

o Crossover 1, which creates a new chromosome by crossing two chromosomes at a single randomly chosen point.

o Crossover 2, which creates a new chromosome by crossing two chromosomes at two randomly chosen points.

- Mutation. There are four forms:

o Shift: creates a new chromosome by shifting all the characters in the string to the left or right by one position. The character leaving the chromosome is inserted on the opposite side, leaving the string the same size.

o Delete: creates a new chromosome by removing a character from a random position. The Levestein distance is increased by the number of deleted elements, to ensure that the gain in distance is not only because the chain is decreased in length, but because an element that was causing the distance to grow was removed. Removing can be taken as a mutation process.

o Insert: creates a new chromosome by inserting a character from the symbol table. The symbol is chosen randomly from the symbol table and the insertion site is chosen randomly from the string, making the string one unit larger. Inserting can be taken as a mutation process.

o Modify: creates a new chromosome by modifying one character with another. The position to be modified is chosen randomly and the character to be inserted is chosen randomly from the symbol table.

The symbol table containing the characters that make up the input strings is used to apply the above functions.

9.3.6 Restrictions

The following are the main constraints that apply to the developed algorithm:

- Regarding execution time: Finding the average string involves dealing with a computationally complex problem called NP-Hard that can take a considerable amount of time. When calculating the average string, from a set of input strings, the processing time required cannot be calculated in advance and varies depending on the length of the strings and their similarity. For this reason the evolutionary algorithm is implemented in the C++ programming language which is a language renowned for its speed of execution .[75]
- Size of the input strings: The use of the C++ language generates a restriction for the inputs of the evolutionary algorithm, which cannot be larger than 12 characters[76] because it causes a memory overflow in the variable that is used as a counter for the combinations of characters that are generated as a search space for the solution.

---

[75] The C++ language allows the generation of native code that runs directly on the computing platform without requiring virtual machines (as in the case of managed code) that slow down the execution of programs.

[76] According to the tests performed, input strings longer than 12 characters generate a variable overflow which leads to an erroneous calculation of the average string.

This situation was identified once the implementation of the evolutionary algorithm was available and the performance tests were carried out; at the end of the document, an alternative is proposed as future work to overcome this important restriction that ended up establishing a limitation in the size of the raw data to be processed.

- Quality of the calculated average string: In the following section it will be shown that the evolutionary algorithm developed to find the average string of a set of given character strings correctly finds the requested average string.

9.3.7 Additional considerations

For the problem of finding the average string using evolutionary programming, no encoding or decoding is required for the inputs and outputs, because these values correspond to strings in natural language. This means that in this particular case the phenotype (input string) is equal to the genotype (internal representation of the input values handled by the algorithm), so the corresponding approach for the implementation of the algorithm is evolutionary programming.

9.3.8 Evaluation of the evolutionary algorithm

To evaluate whether the developed evolutionary algorithm allows to find the average string correctly and in an acceptable time, two different types of evaluations were carried out: first, a contextualised test evaluating the answers provided by the evolutionary algorithm when finding the average string for several input strings corresponding to raw data; second, evaluating the evolutionary algorithm against a realised implementation of an algorithm that calculates the exact average string to compare the response time.

9.3.8.1 Evaluation of the evolutionary algorithm by finding the average chain

This type of test consisted of evaluating the performance of the evolutionary algorithm by finding the average string for several input strings. Different tests of this type were performed with various input strings varying in length, number of inputs and number of symbols. To describe the behaviour of the algorithm, data from three tests are taken.

Test No.1

The input strings for the first test to be analysed are as follows:

- **photocopying Photocopying Photocopying Photocopying Photocopying Photocopying**

These input strings correspond to data collected by 4 different workers for the same field where the exact value (hidden truth) to be collected was the string:

- **Photocopying**

In order to select the number of entries to be used in the tests, recommendations found in the literature in works such as that of Ipeirotis[77] that have studied the relationship between the number of entries and the quality of the results obtained are taken into account. Two general recommendations are made in this regard: the first is to try to have as many entries or contributions as possible, hundreds if possible; and the second is to work with few entries, in the order of 4 to 6 or a few more depending on the task. As collecting a large number of entries for the same task

[77] IPEIROTIS, Panagiotis G., et al. Repeated labeling using multiple noisy labelers. In: *Data Mining and Knowledge Discovery*, 2014, vol. 28, no 2, p. 402.

increases collection costs, the second recommendation is adopted, working with 4 entries.

Table 10-1 gives the average string found by the evolutionary algorithm for ten runs (without varying the input strings), the time of each run and the distance of the average string from the input strings.

Table 9-1. Results evolutionary algorithm test 1

Execution	Medium chain	Weather	Distance
Execution No.1	Photocopying	11,17	4
Execution No.2	Photocopying	11,077	4
Execution No.3	Photocopying	11,123	4
Execution No.4	Photocopying	11,23	4
Execution No.5	Photocopying	11,96	4
Execution No.6	Photocopying	11,114	4
Execution No.7	Photocopying	11,16	4
Execution No.8	Photocopying	11,119	4
Execution No.9	Photocopying	11,31	4
Execution No.10	Photocopying	12,002	4

Source: own elaboration

For this test, the evolutionary algorithm found the average string in approximately 11 seconds. These results are satisfactory because the average string found is equal to the exact value or hidden truth to be found.

Test No.2

In the second test to be analysed, the following input strings were available:

- **fotocoPiado Fotoopiad Foocopado Foticopiad Foticopiadu**

Again the exact value (hidden truth) to be collected was the string:

- **Photocopying**

Table 10-2 shows the results obtained for these input chains.

Table 9-2. Results evolutionary algorithm test 2

Execution	Medium chain	Weather	Distance
Execution No.1	Photocopying	14,134	8
Execution No.2	Photocopying	14,133	8
Execution No.3	Photocopying	13,102	8
Execution No.4	Photocopying	14,182	8
Execution No.5	Photocopying	14,147	8
Execution No.6	Photocopying	14,123	8
Execution No.7	Photocopying	14,164	8
Execution No.8	Photocopying	14,187	8
Execution No.9	Photocopying	11.665	8
Execution No.10	Photocopying	14,139	8

Source: own elaboration

For this test, the evolutionary algorithm found the average string in approximately 14 seconds and despite the fact that the entries had greater dissimilarity, the average string found is equal to the exact value or hidden truth that was to be found.

Test No.3

The third test was carried out with the following input strings:

- **photoCoPying Ftoopiad oocopying Foyicopiadu**

And again the exact value (hidden truth) to be collected was the string:

- Photocopying

Table 10-3 shows the results obtained for these input chains.

Table 9-3. Results evolutionary algorithm test 3

Execution	Medium chain	Weather	Distance
Execution No.1	Fooopado	15,897	15
Execution No.2	Fooopiada	4,321	15
Execution No.3	Focopiada	4,321	15
Execution No.4	Photocopying	4,01	12
Execution No.5	Photocopying	2,917	12
Execution No.6	Photocopying	10,748	14

Source: own elaboration

For this test, the evolutionary algorithm found the average chain but this varied between runs, and the execution time also varied without exceeding 16 seconds. This behaviour is due to the fact that the inputs present much greater dissimilarity than in the two previous tests, as for the time it is still low but this time it presents greater variability.

Similar results were obtained by repeating this type of test with different input strings and varying the degree of dissimilarity in the strings. This allows us to suggest that the evolutionary algorithm correctly finds the average string in an acceptable time, but in order to have more elements to support this approach, a second type of test is carried out in which the evolutionary algorithm is evaluated against a so-called naive algorithm that calculates the exact average string.

9.3.8.2 Evolutionary versus naive algorithm

This test consisted of evaluating the response time of the evolutionary algorithm, using an implementation of an algorithm called naive that calculates the exact mean chain; the implementation of this algorithm is included as a method of the AdmAG class.

It was called a naive algorithm because it calculates the average string but can only be used when the input strings are small in length. For example to calculate the average string corresponding to the following input strings:

- minima Minim Minim Miima Manima

The naive algorithm takes 0.764 seconds, which is a short time for what is said to be a complex computation. But as the length of the strings increases so does the time, for the strings:

- Statute Statute Statute Statute Statute

The naive algorithm takes 397 seconds which is equivalent to 6 minutes, which is already a longer time but still acceptable. But if you increase the length of the string and the symbols that make up the string a bit more, you can have times that are not acceptable. For example, having to calculate the exact average string for entries such as "photocopier" would take approximately 29 days . [78][78]This is why this algorithm that calculates the exact average string is called naïve, as it is not feasible to use it due to the large amount of time required to perform the calculations.

Test No.4

The evaluation consisted of comparing the response times of both algorithms, the evolutionary and the so-called naive, using the same input strings that changed in length (3, 4, 5, 5, 6 and 7 characters) from one run to another, but always retained the same number of symbols that make them up. The input strings were taken from a test corpus[79] doctoral thesis that extensively investigated different ways of finding the average string.

Both algorithms were evaluated when processing input strings consisting of combinations of the following strings varying in length but always consisting of combinations of the symbols (a, b, c, d):

- dac, bdac, caaac, ccaaac, ccaadc, aadaaac

For the same input strings both algorithms always agreed on the average string, the naive one calculating it and the evolutionary one finding it in the search space.

Table 10-4 presents the response times obtained in the tests.

Table 9-4. Response times evolutionary versus naive algorithm

Execution	Response Time Naive Algorithm	Response Time Evolutionary Algorithm
Execution No.1	0,015	0,02
Execution No.2	0,015	0,13
Execution No.3	0,047	0,161
Execution No.4	0,203	0,198
Execution No.5	0,978	0,25
Execution No.6	4,623	0,333
Execution No.7	21,804	0,374

Source: own elaboration

Figure 10-6 allows a comparison of the measured response times in seconds of the algorithms.

Figure 9-6 Response time Naive Algorithm versus Evolutionary Algorithm

[78] If the input string is "Photocopier", you have a symbol table (or alphabet) consisting of: "F", "o", "t", "c", "p", "i", "a", "d", "r" (9 different characters). The naive algorithm has to process ����� possible combinations, i.e. in the order of ���� operations, more precisely 2541865828329, this figure when divided by the number of operations a computer does per second, gives the number of seconds required to process all the combinations. For example if you have a computer that does one million operations per second, then to calculate this input string you would take $(\frac{2541865828329}{10^6} = 2541866 \text{ } seconds = 29$ days).

[79] This work is a benchmark in the problem of calculating the average chain.

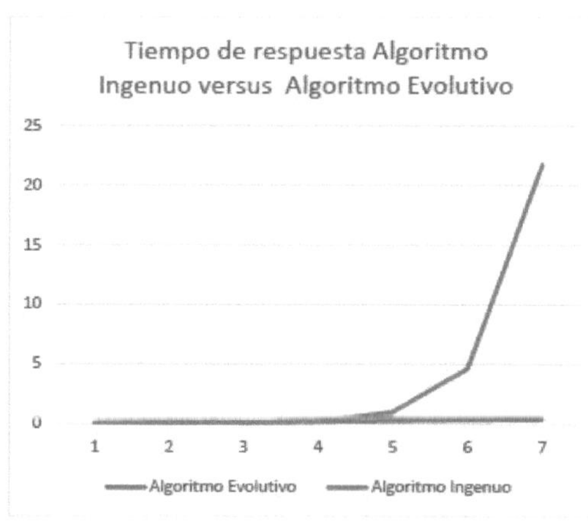

Source: own elaboration

Analysing the behaviour in the response time, it can be stated that for input strings whose length does not exceed 4 characters, the response time is low in both algorithms, with the naive algorithm being faster. But as the input strings increase in length exceeding four characters, the evolutionary algorithm is much faster. The response time of the naive algorithm grows exponentially, while the response time of the evolutionary algorithm grows linearly with a very low slope.

To probe a little into the behaviour of the evolutionary algorithm, several regression models were generated from the available data. After testing a few different types of regression, two models were identified that fit the response delivered by the evolutionary algorithm, as shown in Figure 10-7.

Figure 9-7 Time Trend of Evolutionary Algorithm

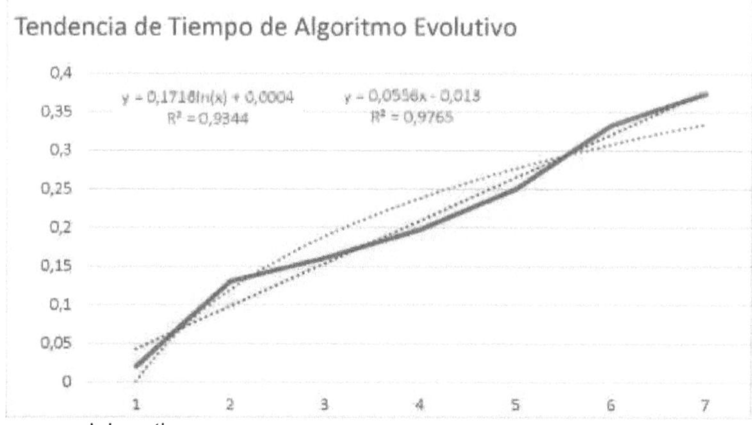

Source: own elaboration

The models shown in Figure 10-7 were obtained from the data in Table 10-4. The first model shown in red corresponds to a logarithmic model and the second model

shown in green corresponds to a linear model. The linear model is the one that best fits the answer given by the evolutionary algorithm according to the coefficient of determination R2 which presents a value of 0.9795.
Similar results were obtained by repeating this type of test with different input strings, varying the length and the number of symbols in them. The results obtained suggest that the evolutionary algorithm can find the average string in a reasonable time because it does not grow exponentially.

CHAPTER 10

10 EVALUATION OF THE METHOD AND FINAL DATA

This section presents the results of the final test of the Crowdsourcing-based data collection method proposed to generate datasets of public procurement processes in the framework of e-government and free access to public information.

10.1 Conduct of the test

10.1.1Objectives of the test

The test carried out was designed to evaluate: first, the process defined in the proposed data collection method, which consists of a series of steps that, when carried out, should generate the desired data set; second, to examine how close the final data generated are to the exact values that were to be collected.

10.1.2 Description of the test

To test the proposed method for data collection, the different steps were carried out with the participation of 8 students of the Systems Engineering programme of the Central Unit of Valle del Cauca (UCEVA), who took the role of Crowdwokers and the author of this paper who carried out the role of Crowdsourcer, the prototype application developed served as a Crowdsourcing platform facilitating the completion of some steps and others, which are not yet supported by the prototype, were carried out manually.

The following is a description of what happened:

Stage I: Initial generation of the dataset

Job 1: Identify

Task 1-1. Select processes: The process starts with the selection of the procurement processes of interest, this task was carried out by the Crowdsourcer (previously registered in the Prototype Application) by performing the following steps:

- Step 1 Indicate transparency portal:
  - Url: https://www.contratos.gov.co/consultas/inicioConsulta.do
- Step 2 Indicate criteria for selecting procurement processes:
  - Entity: Central Unit of the Cauca Valley
  - Contracting modality: Public Tender
  - Date from: 2016-01-01
  - Date until[80] : 2016-12-31
- Step 3: Search the transparency portal using the defined criteria and enter query results:
  - Number of registrations (recruitment processes): 6
  - Number of pages containing results: 1
  - Number of results per page: 50

These steps were performed without reporting any problems using the prototype application. Figure 11-1 and figure 11-2 show the screenshots captured during the steps.

With regard to the criteria defined for the selection of the processes of interest, these were carefully selected so that the number of processes found would be low and it

[80] The end date should always be less than the current date to ensure that the search results do not change as long as the same selection criteria are used.

would be feasible to carry out the test with the 8 participants. The search yielded 6 contracting processes carried out by UCEVA under the public tender modality in 2016, although the figure may seem low these processes total $3,096,896,848,898, Figure 11-3 shows the search performed in SECOP (Electronic Public Procurement System - Colombia) using the defined selection criteria and Figure 11-4 shows the results found.

Figure 10-1. Performing task 1-1 to identify processes

Source: own elaboration

Figure 10-2. Performing task 1-1 query results entry

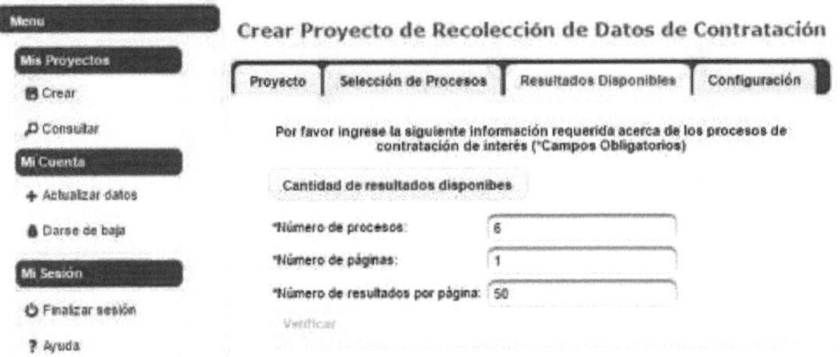

Source: own elaboration

Figure 10-3. Search in Secop

Source: Taken from Secop

Figure 10-4. Search results provided by Secop

Source: Taken from Secop

Task 1-2. Verify selection: The first steps that make up this task were defined in order to be able to verify that the data entered by the Crowdsourcer in the previous task (Task 1-1) were free of errors. These steps are:

- Log in to the transparency portal and indicate whether it is operational.
- Step 5. Perform a search in the transparency portal by entering the defined search criteria and indicate again the results of the query (number of records/processes found, number of pages containing the results, number of results per page).
- Step 6. Verify that the results of the new query match those initially entered.

The testing of these steps was performed manually by the Crowdsourcer, it was not carried out on the prototype application because the requirements associated with this functionality were not selected for implementation.

The last passed defined in the task 1-2:

- Step 7. Create data collection project

It allowed entering the project description data as shown in figure 11-5 and was carried out using the prototype application.

Figure 10-5. Performing task 1-1 project creation

Source: own elaboration

Job 2: Launch

Task 2-1. Configure and publish project: In this task the following steps are performed to configure and publish the project:

- Step 8. Define the basic and intermediate level data or fields to be collected and the number of contributions or responses required.
- Step 9. Define the incentives to be given to motivate and activate the participation of workers in the data collection tasks.
- Step 10. Schedule basic and intermediate data collection tasks for the selected recruitment processes.
- Step 11. Calculate the cost of carrying out the data collection tasks.
- Step 12. Submit cost of data collection project and request confirmation.
- Publish the tasks using another Crowdsourcing platform and recruit workers from the crowd.

The Crowdsourcer using the prototype application carried out the configuration steps (step 8, 9 and 12) as shown in figure 11-6 and figure 11-7, step 11 corresponds to a calculation that is performed by the prototype application.

For the size of the test (6 recruitment processes and 8 Crowdworker users), it is feasible to perform this step manually, but to carry out larger tests, the requirements associated with the task scheduling must be implemented in the prototype application.

The recruitment of workers last step of this task (step 13), is not carried out using another Crowdsourcing platform like for example AMTurk (which have millions of registered Crowdworkers who are available for the realisation of various tasks) because the present research project is an academic work with no budget or funding. In order to recruit the workers, an invitation was extended to several students to participate in this part of the research.

Figure 10-6. Task 2-1 Project configuration realisation

Crear Proyecto de Recolección de Datos de Contratación

Proyecto | Selección de Procesos | Resultados Disponibles | Configuración

Por favor ingrese la siguiente información requerida para configurar el proyecto (*Campos Obligatorios)

Campos a Recolectar por Proceso

*Datos de nivel Básico

*Datos de nivel Intermedio

Redundancia de Respuestas

*Número de contribuciones por campo: 4

Exactitud de los datos

*Nivel de exactitud mínimo requerido: 90 %

Incentivos a Entregar

*Descripción: Las personas que participen en la recolección de datos tendrán acceso al conjunto de datos que sea generado

Source: own elaboration

Figure 10-7. Completion of task 2-1 publication of the Project

Source: own elaboration

Job 3: Describe

Task 3-1. Collect basic level data: The first two steps of this task generate the raw data:

- Step 14. Selecting workers to carry out the basic data collection task
- Access the Urls and pages indicated and identify the records and basic data requested, copy (preferably whenever possible) or transcribe these data.

Step 14 was carried out by the Crowdsourcer and the data collection tasks were assigned to the students who participated by playing the role of Crowdworkers.

Step 15, which is one of the most important steps of the method because this is where the data collection takes place, was carried out by the Crowdworkers using the prototype application. Each Crowdworker was assigned 3 data collection tasks in different processes, figure 11-8 presents the form used in this task; the Crowdworker users were able to do some pre-testing to prepare themselves before carrying out the final exercise.

Annex F contains the raw data collected during the test.

Figure 10-8. Performance of task 3-1 collect data

Recolección de Datos de Contratación

Datos de nivel Básico

Información General del Proceso

*Número de proceso:	129
*Tipo de proceso:	LICITACION PUBLICA
*Estado del proceso:	CELEBRADO
*Clase [Codigo]:	401017
*Clase descripción:	ENFRIAMIENTO
*Cuantia a contratar: $	279500000
*Tipo de contrato:	COMPRAVENTA
*Municipio de ejecución:	TULUA

Información de la Adjudicación del Proceso

*Orden de elegibilidad: (Puntos)	1000
*Nombre o razon social del proponente seleccionado:	PROGEIN LTDA

Información de los Contratos Asociados al Proceso

*Número de contrato:	181
*Estado del contrato:	CELEBRADO
*Cuantia definitiva del contrato: $	332.951.095.00
*Porcentaje de anticipo: %	50
*Ciudad de ubicación del contratista:	BOGOTA
*Plazo de ejecucion del contrato: (dias)	60
*Destinación del gasto:	INVERSIÓN
*Origen de los recursos:	RECURSOS PROPIOS

Source: own elaboration

The last two steps of this task allow the final data to be generated:

- Step 16. Execute quality control procedure based on aggregation of results.
- Step 17. Generate basic data for selected procurement processes

As raw data are potentially noisy, i.e. during collection for some reason errors can be introduced into them that skew their value.

of the exact value to be captured, the method contemplates trying to correct these errors. Step 16, which ended up being the most relevant step of the method, allowed quality control to be performed, which was carried out using the evolutionary algorithm applying the aggregation of results approach.
Finally, given that the intermediate level data collection (task 41) involves the same steps as the basic level data collection (task 3-1), when implementing this part in the prototype application, these tasks were merged into a single form where all the data to be collected were entered. This was possible because all the data of a process is collected from the same source (page), the results obtained for the intermediate level data collection are also presented in the following section.

10.2 Results obtained and discussion

The previous section presented a description of the first part of the test carried out to test the proposed method. In this part, the results obtained showed that by carrying out the different steps that make up the method, the data set of the recruitment processes of interest is generated.
Once the final dataset was available, the second part of the test proceeded, where the objective was to examine how close the final data generated were to the exact values to be collected. For this comparison, a reliable dataset was used that was previously generated for the same procurement processes.
Figure 11-9 and figure 11-10 present together: the raw data collected; the final data generated by the evolutionary algorithm from the raw data (representing a consensus resulting from the aggregation of inputs); the reliable data, which were generated, reviewed and adjusted for errors. All data are associated with the same recruitment process identified by the number 129.

Figure 10-9. Data associated with Procurement Process No.129 - Part 1

DATO CONFIABLE	129	LICITACION PUBLICA	CELEBRADO	401017	ENFRIAMIENTO	279500000	COMPRAVENTA	TULUA	1000	PROGEIN LTDA
DATO FINAL	129	LICITACION PUBLICA	CELEBRADO	401017	ENFRIAMIENTO	279500000	COMPRAVENTA	TULUA	1000	PROGEIN LTDA
DATO BRUTO	129	LICITACION PUBLICA	CONVOCADO	401017	ENFRIAMIENTO	27950000	COMPRAVENTA	BUGA	1000	ROGEIN LTDA
	129	ICITACION PUBLIC	CELEBRADO	401017	ENFRIAMIENTO	279500000	COMPRAVENTA	TULUA	1000	PROEIN LTDA
	129	LICITACION PUBLICA	CELEBRADO	41017	ENFRIAMIENTO	27950000	COMPRAVENTA	TULUA	1000	PROGEIN LTD
	129	LICITACION PUBLICA	CELEBRADO	401017	ENFRIAMIENTO	279500000	COMPRAVENTA	TULUA	1.000	PROGEIN LTDA

Source: own elaboration

Figure 10-10. Data associated with Procurement Process No.129 - Part 2

181	CELEBRADO	332,951,095.00	50	BOGOTA	60	INVERSION	RECURSOS PROPIOS	PROGEIN LTDA	800165236-1	1000	332951095	19459562
181	CELEBRADO	ERROR	50	BOGOTA	60	INVERSION	RECURSOS PROPIOS	PROGEIN LTDA	ERROR	1.000	ERROR	ERROR
181	CELEBRADO	332,951,095	50	BOGOTA	60 Dias	INVERSION	RECURSOS PROPIOS	PROGEIN LTD	800165236-1	1000	33.291.095	19459562
181	CELEBRADO	332,951095	50	BOGOTA	60	INVERSION	RECURSOS PROPIOS	PROGEIN LTDA	800165236-1	1000	332.951.095	19459562
181	CELEBRADO	332,951,095	BOGOTA	BOGOTA	60	INVERSION	RECURSOS PROPIOS	PROGEIN LTDA	800165236	1000	332.951.095	1945962
191	CELEBRADO	332,951,095	50	BOGOTA	60	INVERSIO	RECURSOS PROPIOS	PROGEIN LTDA	800165236-1	1000	332.951.095	19459562

Source: own elaboration

In the figures above, the first row (blue background) corresponds to the reliable data that does not contain errors, the value of each of these fields matches the original value of the field published in Secop (Annex H contains the original information taken from Secop of the processes used for the test). The second row (green background) contains the final data that are generated from processing the raw data corresponding to the final four rows (pink background).

The generated dataset has a total of 23 fields, which are specified in Annex D. The columns correspond to the fields of each dataset and in red are drawn the values that correspond to any errors.

The first column contains the process number, which is an internal identifier assigned by the procuring public entity. For process 129, which is taken as an example, the final dataset presents errors in 4 of the 23 fields, this figure is striking because the objective would be to have no errors at all. But analysing the raw dataset, which is the input to generate the final data, it is observed that it had 15 fields that had at least one error in some of their entries. This means that by applying the quality control mechanism that is performed with the evolutionary algorithm that aggregates the inputs and generates the final data, it was possible to generate correct outputs for 11 of the 15 fields that contained errors.

Similar behaviours and also variations were observed with the data sets of the remaining processes. Figure 11-11 and Figure 11-12 present the data associated with procurement process No.33.

Figure 10-11. Data associated with Procurement Process No.33 - Part 1

RELIABLE DATA	33	PUBLIC TENDER	HELD	921215	SERVICES OF GUARDS	437,267,662	BENEFIT DE SERVICES	TULUA	1.000	TECNISEG DE COLOMBIA LTDA
FINAL DATA	33	PUBLIC TENDER	HELD	921215	ERROR	ERROR	ERROR	TULUA	1.000	ERROR
GROSS DATA	33	PUBLICLY TENDERED	HELD	921215	SERVICES OF GUARDS	437,267,662	OF SERVICES	TULOA	1.000	TECNISEG DE COLOMBIA
	33	PUBLIC TENDER	HELD	92121	SERVICES OF GUARDS	437,267,662	EIN SERVICES	TULUA	100	TECNISEG COLOMBIA LTDA
	33	PUBLIC TENDER	HELD	921215	SERVICES GUARDS	437,267,662	OF SERVICES	TULUA	1.000	TECNISEG DE COLOMBIA LTDA
	33	PUBLIC TENDER	HELD	921215	SERVICES OF GUARDS	437,267	OF SERVICES	TULUA	1.000	TECNISEG DE COLOMBIA LTDA

Source: own elaboration

Figure 10-12. Data associated with Procurement Process No.33 - Part 2

68	HELD	437,182.705.00	ND	ND	263	OPERATION	RESOURCES OWN	COLOMBIA LTDA	830.063.683-1	1000	437.182.705	72.324.818
68	HELD	ERROR	ND	ND	263	ERROR	RESOURCES OWN	ERROR	ERROR	1000	ERROR	ERROR
68	HELD	437,192,705.00	ND	ND	233	FLINCI ON AMI ENT	RESOURCES OWN	COLOMBIA LTDA	830.063.683-1	1000	437.192.705	72.324.819
8	HELD	437,182,705.00	ND	ND	263	OPERATION	RESOURCES OWN	COLOMBIA LTDA	830.063.683	1000	437.182.705	72.324.818
68	HELD	437,182,70	ND	ND	263	OPERATION	RESOURCES OWN	IG FROM COLOMBIA	830.063.683-1	1000	37.182.705	72.324.818
68	HELD	437,182,705.00	ND	ND	263	OPERATION	RESOURCES OWN	COLOMBIA LTDA	830.063.683-1	100	437.182.705	7.232.481

Source: own elaboration

With process 33, the final dataset has errors in 10 of the 23 fields, which is more than in the previous case. When analysing the raw dataset, it was observed that it had 17 fields with at least one error in any of their entries, that is only 1 field more than the previous one, but the errors in the final dataset more than doubled.

To understand what was happening, we examined the other data associated with the remaining processes and were able to identify two situations that caused the evolutionary algorithm to fail to calculate the average string corresponding to the final data:

- The first situation was associated with raw data entries having any word or element longer than 12 characters. This is a constraint of the evolutionary algorithm (discussed in section 10.3.6) that was taken into account and problems were expected with entries corresponding to long words, but it was not contemplated that some fields such as the amount to be contracted or the nit would exceed this limit by including characters such as dots and hyphens. An example of such an entry could be "890.935.513-9" which with 13 characters causes a memory overflow in the algorithm leading in most cases to an error at runtime that does not allow to have an average output string, for this reason in the figures the word ERROR is observed in the fields where this situation occurs.
- The second situation was also related to the restriction on the size of the input strings, which meant that the strings of the same input had to be treated separately. Errors occurred when there were raw entries for a field that consisted of more than two words. If any of the entries had an error due to the lack of one of the words, it meant that the algorithm's input strings were not aligned, producing an average string that showed great dissimilarity with respect to the entries.

Another situation that can lead to not being able to obtain a final data close to or equal to the reliable data is when the raw inputs present a great dissimilarity between them, i.e. they are quite different. This situation did not occur with the data collected for the test but should be considered.

The following figures present the data for the other contracting processes.

Figure 10-13. Data associated with Procurement Process No.159 - Part 1

DATO RELIABLE	159	CALL FOR TENDERS PUBLIC	AWARDED	561015	FURNITURE	464,500.00 0	PURCHASE AND SALE	TULUA	1.000	SPACES AND DESIGNS 21ST CENTURY Ξ.A.Ξ.Ξ
FINAL DATA	159	CALL FOR TENDERS PUBLIC	AWARDED	561015	FURNITURE	FAILURE	PURCHASE AND SALE	TULUA	1000 DOTS	FAILURE
DATO	159	LICITATION	AWARDED	561015	FURNITURE	464.500.00	PURCHASE AND SALE	TULUA	1.000	SPACES AND

GROSS		PUBLIC				0				DESIGNS 21ST CENTURY Ξ.A.Ξ.Ξ
	159	CALL FOR TENDERS PUBLIC	AWARDED	561015	MEBLES	464500000	PURCHASE AND SALE	TULUA	1.000	SPACES AND DESIGNS XXI Ξ.A.A.Ξ
	159	LICITATION PUBLIC	AWARDED	56101	FURNITURE	464.500.00 0	PURCHASE AND SALE	TULIA	1.000	SPACES AND DESIGNS 21ST CENTURY Ξ.A.Ξ.Ξ
	159	LICITATION PUBLIC	AWARDED	601047	FURNITURE	464.500.00 0	PURCHASE AND SALE	TULUA	1.000	SPACES AND DESIGNS 21ST CENTURY Ξ.A.Ξ

Source: own elaboration

Figure 10-14. Data associated with Procurement Process No.159 - part 2

ND	ND	ND	ND	ND	ND	ND	ND	SPACES AND DESIGNS 21ST CENTURY	900.425.814-8	1000	484.000.000	1.130.594.707
ND	ND	ND	ND	ND	ND	ND	ND	FAILURE	FAILURE	1000	FAILURE	FAILURE
ND	ND	ND	ND	ND	ND	ND	ND	SPACES Y DESIGNS 21ST CENTURY	900.425.814-8	1000	464.000.000	113.059.470
ND	ND	ND	ND	ND	ND	ND	ND	SPACES AND DESIGNS XXI Ξ.A.Ξ.A.Ξ	900.425.814-8	1000	64.000.000	1.130.594.707
ND	ND	ND	ND	ND	ND	ND	ND	SPACES Y DESIGNS 21ST CENTURY	900.425.814-8	1000	464.000.000	1.130.584.707
ND	ND	ND	ND	ND	ND	ND	ND	SPACES AND DESIGNS 21ST CENTURY	900.425.814	1000	464.000.000	1.130.584.707

Source: own elaboration

For the previous process No.159, the final dataset shows errors in 6 out of 15 available fields and the raw dataset showed errors in 9 fields, i.e. it is still found that errors are corrected with the processing of the inputs; for this process no contract information was available in Secop, hence the dataset does not have 23 fields as in the previous cases.

Figure 10-15. Data associated with procurement process No.160

DATO RELIABLE	150	PUBLIC TENDER	DISCARDED	4-91716	EQUIPMENT FOR BOXING	7CCCCCCCCCCC	PURCHASE AND SALE	YOUR LUA
FINAL DATA	160	PUBLIC TENDER	DISCARDED	491716	EQUIPMENT FOR	700000000	PURCHASE AND SALE	TULUA
GROSS DATA	160	PUBLIC TENDER	DISCARDED	491716	EQUIPMENT FOR	70000000	PURCHASE AND SALE	TULUA
	150	PUBLIC TENDER	DISCARDED	49176	EQUIPMENT FOR BOXING	700000000	PURCHASE AND SALE	TULUA
	150	PUBLIC TENDER	DISCARDED	491716	EQUIPMENT FOR BOXING	700000000	PURCHASE AND SALE	TULUA
	150	PUBLIC TENDER	DISCARDED	491716	EQUIPMENT FOR BOXING	700000000	PURCHASE AND SALE	TLUA

Source: own elaboration

The contracting process No.160 was discarded, so only a few fields were available in Secop, this time the final dataset generated matched the reliable dataset as all the errors presented (3 in total) could be corrected.

In the last process No.113 the final dataset had errors in 8 out of 23 available fields and the raw dataset had errors in 14 fields, which again shows that errors are still being corrected but the problems mentioned above are still present.

Figure 10-16. Data associated with Procurement Process No.113 - Part 1

DATO RELIABLE	113	CALL FOR TENDERS PUBLIC	HELD	601047	MATERIALS OF PHYSICS OF ENERGY AND ELECTRICITY	700000000	PURCHASE AND SALE	TULUA	1.000	ANALYTIC A ΞΞAΞ
FINAL DATA	113	TENDERING PROCEDURE PUBLIC	HELD	B01047	FAILURE	700000000	PURCHASE AND SALE	TULUA	1.000	FAILURE

DATO GROSS	113	CALL FOR TENDERS PUBLIC	HELD	601047	TE RIALS OF PHYSICS OF ENERGY AND ELECTRICITY	700000000	PURCHASE AND SALE	TULUA	1.000	ANALYTIC A
	113	CALL FOR TENDERS PUBLIC	HELD	601047	MATERIALS OF PHYSICS OF ENERGY AND ELECTRICITY	70000000	PURCHASE AND SALE	TULA	1.000	ANALYTIC A Ξ.A.Ξ
	113	CALL FOR TENDERS PUBLIC	HELD	601047	MATERIALS OF PHYSICS ENERGY AND ELECTRICITY	700000000	PURCHASE AND SALE	TULUA	1.000	ANALYTIC A Ξ.A.Ξ
	113	CALL FOR TENDERS FULICA	HELD	601047	MATERIALS OF PHYSICS OF ENERGY AND ELECTRICITY	700000000	PURCHASE AND SALE	TULUA	1.000	ANALYTIC A Ξ.A.Ξ

Source: own elaboration

Figure 10-17. Data associated with Procurement Process No.113 - Part 2

170	HELD	697,705,084.00	ANALYTICA ΞΞAΞ	MEDELLIN	90	INVESTMENT	RESOURCES OWN	ANALYTIC ASAS	890.935.513-8	1000	687.705.084,00	4.326.602
170	HELD	FAILURE	FAILURE	MEDELLIN	80	INVESTMENT	RESOURCES OWN	FAILURE	FAILURE	1008	FAILURE	FAILURE
170	HELD	697,705,084.00	ANALYTICA	MEDELLIN	90	INVESTMENT	RESOURCES OWN	ANALYTIC AAS	890.935.513-	1000	87.705.084,00	4.326.029
170	HELD	87,705,084.00	ANALYTICA Ξ.A.Ξ	CALLLES	80	INVESTMENT	OWN RESOURCES	ANALYTIC A Ξ.A.Ξ.A.Ξ	890.835.513-8	1008	68.770.504,00	43.268.029
170	HELD	697,705,084.00	ANALYTICA Ξ.A.Ξ	MEDELLIN	80	INVESTMENT	OWN RESOURCES	ANALYTIC A Ξ.A.Ξ.A.Ξ	880.35.513-8	1008	687.705.084,00	43.268.029
70	HELD	697,705,084.00	ANALYTICA Ξ.A	MEDELLIN	80	INVERDION	RESOURCES OWN	ANALYTIC A Ξ.A.Ξ	890.835.513-8	10	687.705.084,00	43.266.029

Source: own elaboration

Table 20 summarises the results obtained for the test where all final data generated for the recruitment processes have fewer fields with errors than the raw data sets, which always have errors.

Table 10-1 Summary of results obtained

Process	No. of Raw Fields with Errors	No. of Fields in Endgame with Errors	No. Fields Corrected
Process No.129	15 of 23	4 of 23	11
Process No.33	17 of 23	10 of 23	7
Process No.113	14 of 23	8 of 23	6
Process No.159	9 of 15	6 of 15	3
Process No.160	3 of 8	0 of 8	3

Finally, it should be noted that the incorporation of the human element in the data collection process made it possible to generate the dataset but with some errors that in certain cases could be corrected using the evolutionary algorithm that acted as a quality control mechanism.

CONCLUSIONS

In the scientific literature there are a number of works where Crowdsourcing has been used as a method of data collection, for different situations and areas of application with good results. Surprisingly, none were found that could be used as a general guideline when defining the method. Different loose elements were collected from various works and were subsequently used in the definition of the proposed method together with other elements that were found following the procedure defined for its elaboration.

The proposed data collection method was tested and the results obtained showed that by carrying out the different steps specified, the data set of the recruitment processes of interest is generated. When examining how close the final data generated is to the exact values to be collected, it was found that the incorporation of the human element in the data collection process introduced some errors that in certain cases could be corrected using the evolutionary algorithm that acted as a quality control mechanism.

Among the errors identified, it was found that some are generated: when transcribing the requested data, this type of error introduces erroneous characters in any position of the input string; when copying the requested data from the source page, this error may delete characters at the beginning or end of the string or even incorporate characters that are not part of the original data to be copied; copying or transcribing a totally erroneous value. It was also identified that fewer errors occur when the GUI uses inputs with closed controls, such as a selection list, but these types of controls can only be used when the different possible input values are known.

As for the evolutionary algorithm, it was designed with the purpose of using it to generate final data from the raw data collected through Crowdsourcing, for this reason it implements the aggregation of results approach to find the average string it finds in a suitable way. But by using it to aggregate redundant inputs, the quality of the response depends on the input strings, which must not have large deviations from the exact value to be captured. This is a restriction associated with the use of the algorithm and not with the algorithm as such, which does present a restriction on the size of the input strings, which cannot exceed 12 characters, because this situation means that the average string cannot be calculated. As for the time required by this algorithm to calculate the average string, it was found that it does not grow exponentially with respect to the size of the string, which in practice means that the average string can be found in a few seconds.

The data published in Secop is a valuable source of information on public procurement that could be used to generate datasets that may be even more comprehensive than those provided by EU countries, where procurement systems only publish a few data for each procurement process carried out.

Finally, it can be said that the results of the work carried out provide an answer to the research question on how to generate datasets of public procurement processes from unstructured information published on the Web using crowdsourcing as a method of data collection. Bearing in mind that the results obtained allow us to identify the presence of problems associated with the quality that these datasets may have, the work to follow should raise questions on how to improve the quality of these datasets.

RECOMMENDATIONS AND FUTURE WORK

Here are some recommendations and some of the future work to be developed:

- In order to overcome the restriction on the size of the input strings in the evolutionary algorithm, it is proposed as future work to carry out a new implementation of this algorithm using a non-typed language such as Racket (Scheme) that does not impose limitations on the size of the memory allocated to the variables, so that the length of the input strings could be extended.
- Test the performance of the developed evolutionary algorithm for finding the average chain against other recognised state-of-the-art algorithms.
- Conduct real scenario testing to generate large datasets and be able to come up with generalisable results.
- Develop a new, more robust quality control mechanism that combines various techniques such as pre-selection, gold data, and reputation, also taking advantage of the evolutionary algorithm developed.
- Develop a new method of recruitment data collection but combining information retrieval and extraction techniques that are complemented by crowdsourcing to take advantage of the benefits of each.
- Develop specialised validators according to the nature of the data to be captured, to be used at the time of data entry to avoid or guarantee certain conditions and/or formatting in the input strings.
- Develop a crowdsourcing-based method for assessing red flags (indicators of possible corruption) in datasets and use all this data to generate analytical models that can be used to study what happens in public procurement.

BIBLIOGRAPHY

AMSTERDAMER, Yael, GROSSMAN, Yael, MILO, Tova, and SENELLART, Pierre. Crowd mining. In: International Conference on Management of Data. s.l: ACM, 2013, p.241.

BARDHAN, Pranab. Corruption and development: a review of issues. In: Journal of Economic Literature, 1997, vol.35, no.3, p.1320.

BARR, Joshua, PINILLA, Edgar, and FINKE, Jorge. A Legal Perspective on the Use of Models in the Fight Against Corruption, In: South Carolina Journal of International Law and Business. 2012, vol.8, no.2, p.266.

BENATALLAH, A., et al. Quality Control in Crowdsourcing Systems: Issues and Directions. In: IEEE Internet Computing, 2013, vol.17, no.2, p.76.

BERLOT, Carlo, JAEGER, Paul, and GRIMES, Justin. Using ICTs to create a culture of transparency: E-government and social media as openness and anticorruption tools for societies. In: Government Information Quarterly, 2010, vol.27, no.3, p.264.

Crowd-Sourcing Transparency: ICTs, Social Media, and Government Transparency Initiatives. In: 11th Annual International Conference on Digital Government Research, Puebla: ACM, 2010, p.53.

BRABHAM, Darem. Crowdsourcing as a model for problem solving: An introduction and cases. In: Convergence: The International Journal of Research into New Media Technologies, 2008, vol.14, no.1, p.75-90.

BRUNETTI, Aymo, and WEDER, Beatrice. A Free Press is Bad News for Corruption. In: Journal of Public Economics, 2003, vol.87, no.7, p.1801-1824.

BUSCAGLIA, Edgardo, and DIJK, Van. Controlling Organized Crime and Corruption in the Public Sector. In: Forum on Crime and Society, 2004, vol.3, no.1, p.3.

CARVAJAL, Raul. Large-scale corruption: definition, causes and cures. In: Systemic Practice and Action Research, 1999, vol.12, no.4, p.335.

CEES, Van der Eijk and Mark Franklin. Elections and voters: Assessing electoral democracy. London: Palgrave Macmillan, 2009, p.213.

CHEN, Guoying, et al. The Effectiveness and Limitations of Internet AntiCorruption A Perspective of Institutional Analysis. In: International Conference on Management of e-Commerce and e-Government. Beijing: IEEE, 2012, p.85.

CLAD. Ibero-American Charter on e-Government. [online]. Santiago de Chile: November 2007 [accessed 8 February 2016]. Available online: CRIADO, Ignacio. and RAMILO, Carmen. Towards an Integrated Vision of Electronic Government. In: Revista Vasca de Economía. 2003, vol.3, no.54, p.194.

CRIADO, Ignacio, and COROJAN, Ana. Can ICTs change transparency, anti-corruption and accountability in Latin American governments: a comparative approach to Central American countries. In: XIV Encuentro de Lationoamericanistas Españoles 200 años de Iberoamérica (1810-2010). Santiago de Compostela: Universidad de Santiago de Compostela, 2010, p. 2037.

DE GRAFF, Gjalt, and HUBERTS, Leo. Portraying the nature of corruption using an explorative case study design. In: Public administration review, 2008, vol.68, no.4, p.640.

DiRienzo, Cassandra, et al. Corruption and the Role of Information. In: Journal of International Business Studies, 2007, vol.38, no.1, p.320-332.

DUNLEAVY, Patrick, et al. Digital Era Governance: IT Corporations, the State and E-government. Oxford: Oxford University Press. 2006. p.28.
ERICKSON, Thomas. Geocentric Crowdsourcing and Smarter Cities: Enabling Urban Intelligencein Cities and Regions. In: 1st International conference on ubiquitous computing. Copenhagen: ACM, 2010, p.4.
ESTELLÉS, Enrique, GONZÁLEZ, LADRÓN-DE-GUEVARA. Towards an integrated crowdsourcing definition. In: Journal of Information Science, 2012, vol.38, no.2, p.189-200.
FAJARDO, Luis. La Corrupción Heredada: Pasado Colonial, Sistema Legal y Desarrollo Económico en Colombia. In: Revista de estudios sociales Universidad de los Andes. 2002, vol.1, no.12, p.22.
FAN, y., ZHANG, z., and YUE, q. E-government, Transparency and Anti-corruption: How can E-government reduce corruption effectively in China?. In: *International* Conference on Management of e-Commerce and e-Government. s.l: IEEE, 2009. p.101.
FAN, Yujian, ZHANG, Zengtian, and YUE, Qingchun. E-government, Transparency and Anti-corruption. How can E-government reduce corruption effectively in China? In: 9th International Conference on Management of e-Commerce and eGovernment, Nanchang: IEEE, 2009, p.101.
FASEKAS, Mihály, TÓTH, János, and KING, Peter. Anatomy of grand corruption: A composite corruption risk index based on objective data. Budapest: Corruption Research Center Budapest, 2013. p. 9.
FASEKAS, Mihály, TÓTH, János, and KING, Peter: Corruption manual for beginners: Corruption techniques in public procurement with examples from Hungary. Budapest: Corruption Research Center Budapest, 2013. p.10.
FLORINI, Ann. The Right to Know: Transparency for an Open World. s.l: Columbia University Press, 2007, 376 pp.
GASCÓ, Mila. The role of institutions in the development of e-government in Latin America: some reflections. In: Revista Chilena de Administración Pública, 2009, vol.1, no.14, p.37.
GOKHALE, Chaitanya, et al. Corleone: Hands-off crowdsourcing for entity matching. In: ACM SIGMOD International Conference on Management of Data. Snowbird: ACM, 2014, p.601.
GOMEZ, Iván, and VILLA, Fredy. *Control al Control de la Contratación Riesgos, Desaciertos y Posibilidades.* Bogotá: Auditoría General de la Republica. 2010. p.3.
HEEKS, Richard. Information Systems for Public Management: Information Technology and Public Sector Corruption. Manchester: IDPM-University of Manchester, 2000, p.1-15.
HOSSEINI, M., PHALP, K., TAYLOR, J. and R. Ali. The Four Pillars of Crowdsourcing: A Reference Model, In: 8th International Conference on Research Challenges in Information Science. Marrakesh: IEEE, 2014, p.1-12.
HOWE, Jeff. The rise of crowdsourcing. In: Wired Magazing, 2006, vol.14, no.6, p.61. http://www.wired.com/2006/06/crowds/
IPEIROTIS, Panagiotis G., et al. Repeated labeling using multiple noisy labelers. Data Mining and Knowledge Discovery, 2014, vol. 28, no 2, p. 402-441.
KAZAI, Gabriella, KAMPS, Jaap, and MILIC-FRAYLING, Natasa. Worker types and

personality traits in crowdsourcing relevance labels. In: 20th International conference on Information and knowledge management. Glasgow: ACM, 2011, p.1941.

KOHONEN, Teuvo. Median strings. Pattern Recognition Letters, 1985, vol. 3, no 5, p. 309-313.

LARMOUR, Peter, and WOLANIN, Nick. Corruption & Anti-corruption. Canberra: Asia Pacific Press, 2001, 262 pp.

LEI, Duan, SATOSHI, Oyama, HARUNHIKO, Sato, and MASAHITO, Kurihara. Separate or joint? Estimation of multiple labels from crowdsourced annotations. In: Expert Systems with Applications, 2014, vol.41, no.13, p.5723.

MAHMOOD, Rumel. Can information and communication technology help reduce corruption? How so and why not: Two case Studies from South Asia. In: Perspectives on Global Development and Technology, 2006, vol.3, no.3, p.347.

MATHEUS, Ricardo, RIBERIRO, Manuella, VAZ, José, and SOUZA, Cesar. Internet Use for Social Control and Participation: What Are Local Governments Doing in Latin America? In: 4th International Conference on Theory and Practice of Electronic Governance. Beijing: ACM, 2010, p.333.

MCCULLAGH, Karen. E-democracy: potential for political revolution? In: International Journal of Law and Information Technology, 2003, vol.11, no.2, p.149-161.

MURUGESAN, San. Understanding Web 2.0. In: IT Professional, 2007, vol.9, no.4, p.34.

NDOU, Valentina. E-Government for Developing Countries: Opportunities and Challenges. In: The Electronic Journal of Information Systems in Developing Countries (EJISDC), 2005, vol.18, no.1, p.15.

NICHOLLS, Collins, et al. Corruption and misuse of public office. Oxford: Oxford University Press, 2006, p.20.

O'REILLY, Tim. Web 2.0: Compact Definition? [online] October 2005. [accessed 9 February 2016]. Available online: http://radar.oreilly.com/2005/10/web-20-compact-definition.html

OLESON, David, et al. Programmatic gold: Targeted and scalable quality assurance in crowdsourcing. In: 25th AAAI Conference on Artificial Intelligence. San Francisco: Association for the Advancement of Artificial Intelligence, 2011, p.43.

ORIOL, P, Joan. New technologies, democracy and efficiency of political institutions. In: Revista de Sociología. 2005, vol.78, p.151.

PAOLACCI, Gabriele, CHANDLER, Jesse, and IPEIROTIS, Panagiotis. Running Experiments on Amazon Mechanical Turk. In: Judgment and Decision Making, 2010, vol.5, no.5, p.411.

PAPADOPOULOU, Chrysaida-Aliki, and GIAOUTZI, Maria. Crowdsourcing as a Tool for Knowledge Acquisition in Spatial Planning. In: Future Internet. 2014, vol.6, no.1, p109.

PORTER, Joshua. Designing for the Social Web. s.l: New Riders Press, 2008, p.5.

QUINN, Alexander, and BEDERSON, Benjamin. Human computation: a survey and taxonomy of a growing field. In: Conference on Human Factors in Computing Systems. Vancouver: ACM, 2011, p.1403-1412.

REILLEY, Katherine, and ECHEBERRIA, Raúl. The Role of the Citizen and CSOs in e-Government: A Study of e-Government in Eight Latin American and Caribbean Countries. [online]. January 2003 [accessed 8 February 2016]. Available online:

http://lac.derechos.apc.org/investigacion/e- gobierno.pdf
RELLY, J. E., and SABHARWAL, Meghna. Perceptions of transparency of government policymaking: A cross-national study. Government Information Quarterly, 2009, vol.26, no.1, p.148.
ROBERTS, Alasdair. Blacked out: Government secrecy in the information age. New York: Cambridge University Press, 2006, p.322.
ROSE-ACKERMAN, Susan. The Role of The World Bank in Controlling Corruption. In: Law & Policy in International Business, 2007, vol.29, no.1, p.29.
ROSS, J., et al. Who are the crowdworkers: shifting demographics in mechanical turk. In: 28th International Conference on Human Factors in Computing Systems. Atlanta: ACM, 2010, p.2863.
SANTOS, L. A. Accountability, access to information and participation and government decision-making. In: X International Congress of CLAD on State Reform and Public Administration, Santiago, Chile: CLAD, 2005, p.18.
SCHENDLER, Andreas, Plattner, Mark, and Diamond, Larrry. The self restraining state: power and accountability in new democracies. s.l : Lynne Riener Publishers, 1999. p.395.
SHIM, Dong, and EOM, Tae. Egovernment and anti-corruption: Empirical analysis of international data. In: International Journal of Public Administration, 2008, vol.31, no.3, p.298-316.
SHIM, Dong, EON, Tae, HONDEGHEM, A. and PERRY, James. Anticorruption effects of information communication and technology (ICT) and social capital. In: International Review of Administrative Sciences, 2009, vol.75, no.1, p.99.
SHULER, Jhon, JAEGER, Paul, and BERLOT, Carlo. Implications of harmonizing egovernment principles and the Federal Depository Library Program (FDLP). In: Government Information Quarterly, 2010, vol.27, no.1, p.9.
SMITH, Robert, and BERTOZZI, Mark. Principals and agents: An explanatory model for public budgeting. Journal of Public Budgeting. In: Accounting & a Financial Management. 1998, vol.10, no 3, p.325.
STEVENS, Mile, and GANASELVAM, Shiro. The World Bank and Governance. In: IDS Bulletin, 1995, vol.26, no.2, p.97.
STIGLIZ, Joseph. Information and the change in the paradigm in economics. In: American Economic Review, 2020, vol.92, no.3, p.460.
SUN, Chong, RAMPALLI, Narasimhan, YANG, Frank, and DOAN, Anhai. Chimera: Large-scale classification using machine learning, rules, and crowdsourcing. In: Very Large Database Endowment. 2014, vol.7, no.13, p.1529.
TANZI, Vito. Corruption Around the World Causes, Consequences Scope, and Cures. [online] May 1998. [accessed 9 February 2016]. Available online: http://www.imf.org/external/pubs/ft/wp/wp9863.pdf
TESORO, José. Transparencia y tecnologías de gestión: la viabilidad de sistemas con transparencia en la administración pública. In: Estado y Sociedad, vol1, no.33, p.33.
TSE-TUNG, M. On Contradiction [online]. August 1937 [accessed 8 February 2016]. Available online:
http://www.marxists.org/reference/archive/mao/selected-works/volume-1/mswv117.htm

TWEDDLE, J., L. ROBINSON, D., POCOCk, M.J.O., ROY, H.E. Guide to citizen science: developing, implementing and evaluating citizen science to study biodiversity and the environment in the UK. [online] S.l: Natural History Museum and NERC Centre for Ecology & Hydrology for UK-EOF, 2012, 36 p. [accessed 9 February 2016]. Available online at: http://www.nhm.ac.uk/content/dam/nhmwww/take-part/Citizenscience/citizen-science-guide.pdf

UNITED NATIONS. United Nations Convention Against Corruption. [online] New York: 2004. [accessed 09 February 2016]. Available online: http://www.unodc.org/pdf/corruption/publicationsunodc convention-s.pdf

VALCERDE, M. (2004). Transparency, access to information and accountability: Experiences from the European Union and Mexico. In: IX Congreso Internacional del CLAD sobre la reforma del estado y la administración pública. Madrid: Clad, 2004, p.8.

WORLD BANK. Helping Countries Combat Corruption: The Role of the World Bank. [online]. September 1997. [accessed 9 February 2016]. Available online: http://www1.worldbank.org/publicsector/anticorrupt/corruptn/cor02.htm

YAN, Tingxin, KUMAR, Vikas, and GANESAN, Deepak. Crowdsearch: Exploiting crowds for accurate real-time image search on mobile phones. In: 8th international conference on Mobile systems, applications, and services. San Francisco: ACM, 2010, p.77.

ZHANG, Chen, CHEN, Lei, JAGADISH, H.V., and CAO, Chen. Reducing uncertainty of schema matching via crowdsourcing. In: Very Large Database Endowment. 2013, vol.6, no.9, p.757.

Printed by Books on Demand GmbH, Norderstedt / Germany